Congress on Industrial and Applied Life Sciences & Mathematics (Nature-Math 2016)

Journal of Physics: Conference Series Volume 818

Yogyakarta, Indonesia
5 - 6 November 2016

Editors:

Lumban Gaol

ISBN: 978-1-5108-3979-3
ISSN: 1742-6588

Printed from e-media with permission by:

Curran Associates, Inc.
57 Morehouse Lane
Red Hook, NY 12571

Some format issues inherent in the e-media version may also appear in this print version.

Printed by Curran Associates, Inc. (2017)

Institute of Physics
Dirac House, Temple Back
Bristol BS1 6BE UK

Phone: 44 1 17 929 7481
Fax: 44 1 17 920 0979

techtracking@iop.org

Additional copies of this publication are available from:

Curran Associates, Inc.
57 Morehouse Lane
Red Hook, NY 12571 USA
Phone: 845-758-0400
Fax: 845-758-2633
Email: curran@proceedings.com
Web: www.proceedings.com

Table of contents

Volume 818

2016 Congress on Industrial and Applied Life Sciences & Mathematics (Nature-Math 2016)

5–6 November 2016, Yogyakarta, Indonesia

Accepted papers received: 1 March 2017
Published online: 23 March 2017

Preface

Papers

IOP Conf. Series: Journal of Physics: Conf. Series **818** (2017) 011001 doi:10.1088/1742-6596/818/1/011001

The 2016 Congress on Industrial and Applied Life Sciences & Mathematics (The Nature-Math 2016)

The Hotel EastParc Yogyakarta, Indonesia
05 - 06 November 2016

Preface

The 2016 Congress on Industrial and Applied Life Sciences & Mathematics (The Nature-Math 2016) alerady held at the Hotel EastParc Yogyakarta, Indonesia, during 05 - 06 November 2016.

The Nature-Math 2016 is designed to address one of the most challenging and urgent scientific and social demand that face global societies in the twenty-first century: life sciences for social good.
The structure of the conference features seven main themes of life sciences and mathematics: Environment, Food and Nutrition, Communication Disorders as human science, the use of Information Technology, Numerical Analysis and Computational Mathematics, Applied and Industrial Mathematics, and Mathematics Education.
This conference intends to bring together researchers, practitioners, researchers, and scholars from a wide range of disciplines who have a shared interests in the above-mentioned themes. We welcome contributions addressing any of the main themes of the conference.

The theme of The Nature-Math 2016 " The Industrial Applications of Research Result of Life Sciences & Mathematics"

This year, we received 29 papers and after rigorous review, 04 papers were accepted.

The participants come from 04 countries. There are 1 (one) Paralell Session and 3 Keynote Speakers.

It is an honour to present this volume of Journal of Physics: Conference Series (JPCS) and we deeply thank the authors for their enthusiastic and high-grade contribution.

We would like to express our sincere gratitude to all in the Technical Program Committee who have reviewed the papers and developed a very interesting Conference Program as well as the invited and plenary speakers.

Finally, we would like to thank the conference chairmen, the members of the steering committee, the organizing committee, the organizing secretariat and the financial support from the conference sponsors that allowed the success of The Nature-Math 2016.

The Editor of the The Nature-Math 2016 Proceedings Editor
Dr. Ford Lumban Gaol
Bina Nusantara University

IOP Conf. Series: Journal of Physics: Conf. Series **818** (2017) 011002 doi:10.1088/1742-6596/818/1/011002

The 2016 Congress on Industrial and Applied Life Sciences & Mathematics (The Nature-Math 2016)

The Hotel EastParc Yogyakarta, Indonesia
05 - 06 November 2016

1. T.C. Chang

Department of Geography
National University of Singapore

Art and Creative Placemaking in Singapore

Creative placemaking is a process where different sectors of society (state, arts organisation and community) collaborate in cultural activities that deepen their relations with each other and also the place in which they operate and live in. As a panacea to the woes of poorly planned spaces and apathetic people, it is supposed to address social exclusion and generate feelings of belonging even as the environment is artfully and culturally enhanced. Adopting a creative placemaking framework, this presentation considers the phenomenon of Void Deck Galleries (VDG) in Singapore. With more spaces in public housing environments given over to art in Singapore, it explores the relationships between the state, an arts organisations and local communities that have formed around the VDGs, along with the missed opportunities and looming challenges in urban placemaking. Data from three void decks reveal a pro-active stance by the state, the coordinating function of the arts organisation, and the enthusiastic although largely passive role that local communities bring to bear on the painted walls of Singapore.

IOP Conf. Series: Journal of Physics: Conf. Series **818** (2017) 011002 doi:10.1088/1742-6596/818/1/011002

Biography

T.C. Chang is an Associate Professor in Geography at the National University of Singapore. His research interests include urban development issues, arts and culture, and Asian tourism. He is the co-editor of two books on Asian tourism namely, Asia on Tour. Exploring the Rise of Asian Tourism (2009; Routledge) and Interconnected Worlds: Tourism in Southeast Asia (2001; Pergamon). He was also Assistant and subsequently Vice Dean of the Faculty of Arts and Social Sciences in charge of External Relations and Student Life (2008-2015). In 2016, he is currently on sabbatical where he was a visiting associate at the Chinese University of Hong Kong and Surrey University (UK).

2. Professor Datuk Dr. Sufean Hussin , Ph.D (Oregon, USA), Master Education., B.Sc & Ed.(Hons)

Dept. of Educational Management, Planning & Policy
Faculty of Education, University of Malaya
Kuala Lumpur, MALAYSIA

Email address: drsufean@um.edu.my
Fax number: +603 7967 5010
Handphone: +6011 2629 0982

Professor Datuk Dr Sufean Hussin is a senior special grade professor and has served the University of Malaya for more than 32 years.

Achievements:

1. He is one of the distinguished scholars in the field of education in the world and in Malaysia, specializing and researching in the areas of university governance and management, educational leadership, and education policy.
2. He has been honoured as "Top 2000 Education Intellectuals of the 21st Century" by the International Biography Centre (IBC Cambridge) and listed in the biography annal "Who Is Who in the World" (New York 2011, 2012).
3. He has been a research consultant to Ministry of Education on several occassions and also to several corporations and associations. His most notable consultancy work was with the MoE, 2012-2014, with a grant of RM6.9 million, doing a major research project on teacher employment and services policy.
4. He has written more than twenty books, among the notable ones are:
    - Pendidikan di Malaysia (1996);
    - Dasar Modal Insan: Dasar 60:40 Dalam Sains & Teknologi
    - Demokrasi Pendidikan: Dilema Sekolah Kecil dan Sekolah Berpusat (2013);
    - University Governance: Trends and Models (2012);
    - Memacu Puncak Ilmu: Autonomi Universiti Merencana Pembangunan (2008).
5. He has supervised research works and theses of more than 120 masters students and 12 PhD students.
6. He is listed as one of the United Nation's experts in field of education and social science, 2013-2015.
7. For his excellent service, he was awarded the Datukship medal by the Malacca Governor in September 2013, and awarded the Excellent Service award by the University of Malaya for three times.

3. Mohamed Sarrab, Ph.D

Communication and Information Research Center,
Sultan Qaboos University, Muscat, Sultanate of Oman

IOP Conf. Series: Journal of Physics: Conf. Series **818** (2017) 011002 doi:10.1088/1742-6596/818/1/011002

Human Behavior Consideration in Mobile Learning Development and Adoption

The growth of (smart) mobile devices usage, increased acceptance of E-learning, the availability of high reliability mobile networks and need for flexibility in learning have resulted in the growth of mobile learning (M-learning). This has led to a tremendous interest in the acceptance behaviors related to M-learning users among the information systems researchers. The last decade has seen emergence of a new economy called Knowledge Economy, a fusion of globalization and Information and Communication Technology (ICT). M-learning provides a new learning channel in which learners can access content and just in time information as required irrespective of the time and location. Even though M-learning is fast evolving in many different societies, research addressing the driving factors of M-learning development and adoption is in short supply.

Educationally advanced countries such as South Korea, USA, Japan, Singapore, European Union and Australia are using mobile devices in the educational sectors. These countries have been found promoting M-Learning as a matter of national policies. They have accommodated E-Learning and M-Learning in their traditional learning systems. However most of the other countries are still out of race and facing number of challenges in M-Learning adoption.

This talk focuses on the driving factors as human behavior consideration in mobile learning development and adoption and the society perceptions and willingness towards M-learning. Some factors of perceived innovative characteristics, such as ease of use, usefulness, enjoyment, suitability, social, and economic were found to have more influence on development and adoption of M-learning which help to facilitate and promote future research. Empirically, I draw on results from research project that investigate the development, adoption, and dissemination of M-learning. The aim of this speech is to provide such insights based on empirical study and to raise adequate inquiries for debate and research.

Biography

Dr. Mohamed Sarrab is currently working as a researcher at CIRC (Communication and Information Research Center), Sultan Qaboos University, Muscat, Sultanate of Oman. He obtained Ph.D. degree in computer science from De Montfort University, UK. M.Sc. degree in computer science and information technology from VSB technical university of Ostrava, Czech Republic and B.Sc. degree in computer science from Al Zawia University, Libya.

His research interests are in areas of Computer Security, Software Engineering, E-learning, Mobile learning (M-learning) and Mobile Cloud Computing. He is also interesting in mobile application security, in particular Access Control and Policy-Based System Management, Runtime Verification and Information Flow Control, viz. Software Systems where security requirements are managed using loosely coupled components that enforce high-level security requirements. He believed that any such system must be based on a sound and formal footing to provide the level of assurance needed in critical applications, whilst remaining practical and usable.

Peer review statement

All papers published in this volume of *Journal of Physics: Conference Series* have been peer reviewed through processes administered by the proceedings Editors. Reviews were conducted by expert referees to the professional and scientific standards expected of a proceedings journal published by IOP Publishing.

Development of methods for analysis of knee articular cartilage degeneration by magnetic resonance imaging data

Artjoms Suponenkovs [1], Aleksandrs Glazs [1], Ardis Platkajis [2]

[1] Riga Technical University, Faculty of computer science and information technology, Setas str. 1, LV-1048, Riga, Latvia

[2] Medical Academy of Latvia, Dzirciema str. 16, LV-1007, Riga, Latvia

E-Mail: artjoke@inbox.lv(Artjoms Suponenkovs), glaz@egle.cs.rtu.lv(Aleksandrs Glazs)

Keywords: early diagnosis of osteoarthritis, magnetic resonance imaging, medical imaging, knee joint, cartilage, image segmentation, relaxation time

Abstract. The aim of this paper is to describe the new methods for analyzing knee articular cartilage degeneration. The most important aspects regarding research about magnetic resonance imaging, knee joint anatomy, stages of knee osteoarthritis, medical image segmentation and relaxation times calculation. This paper proposes new methods for relaxation times calculation and medical image segmentation. The experimental part describes the most important aspect regarding analysing of articular cartilage relaxation times changing. This part contains experimental results, which show the codependence between relaxation times and organic structure. These experimental results and proposed methods can be helpful for early osteoarthritis diagnostics.

1. Introduction

The analysis of knee articular cartilage degeneration by magnetic resonance imaging data is very important due to decreasing human physical activity and increasing number of people with osteoarthritis(OA) [5]. Every tenth American adult has OA, this causes economic losses. Therefore, timely OA diagnosis is very significant. While the patient is still in the first stage of their OA, it is possible to completely the disease. Magnetic resonance imaging (MRI) is the most effective noninvasive technique for analyzing and displaying articular cartilage damage. Thus, this paper describes new methods for knee cartilage early osteoarthritis diagnostic by magnetic resonance imaging data.

2. Problem statement

An early osteoarthritis diagnosis is very important because it gives us the possibility to completely cure osteoarthritis. The knee articular cartilage changes are recoverable at the early stage of the OA. Unfortunately, doctors can't see OA at an early stage of the disease, this makes early diagnosis of OA even more important. Another important problem is cartilage segmentation [6-9]. By solving the problem of cartilage segmentation, we will be able to solve cartilage degeneration analysation problem. The cartilage degeneration analysation is useful for automatic cartilage degeneration analysation. This paper will discuss the following problems: OA early stage displaying, cartilage segmentation, cartilage degeneration analysation. In the paper, various different methods are proposed for solving these problems.

3. Proposed approach

Osteoarthritis has 5 grades (figure 1) [12, 13]. Doctors are unable to see the first grade of the OA using simple grayscale MRI images. But the knee articular cartilage changes are recoverable at the first grade of the OA. So, OA first grade displaying is a very important problem.

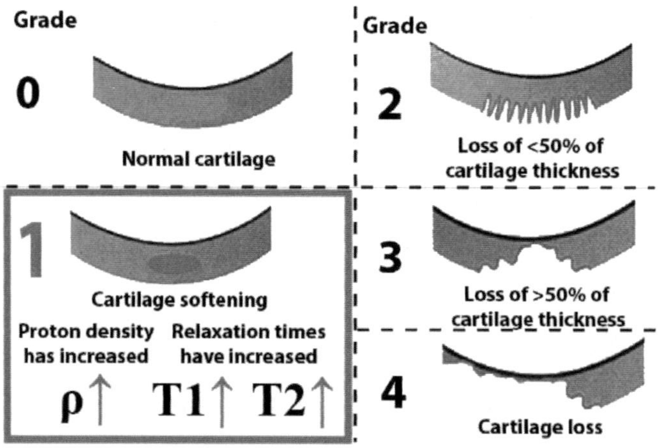

Figure 1. Outerbridge classification.

The first grade of OA has the following features: increase in the proton density and increase in relaxation times [10], where proton density – is number of hydrogen resonating proton per unit volume and T1/T2 relaxation times – is the recovery times of the protons spin magnetization after RF exposure. Thus, this paper describes methods for relaxation times calculation using one MRI image, two MRI images and multiple MRI images. The result of calculation (relaxation times) is shown by colorful images. This colorful image is useful for displaying relaxation time changes. Image pixel colors (eg. we can use Hue Saturation Value color system) are displayed according to calculated relaxation times.

After relaxation time calculation it is important to make cartilage segmentation. If cartilage segmentation has been done successfully, you will be able to make a cartilage analyzation by relaxation time. In the experimental part has been used interactive segmentation method. For cartilage analysing we can use the dispersional analysis method and histogram analysis method. Common flowchart is shown in figure (figure 2).

Simple MRI image pixels value is dependent on MRI signal (S). But the MRI signal is dependent on relaxation times T2, T1 as well as proton density ρ [2]. We can change relaxation times and proton density influence on MRI signal. The dependence between signals and relaxation times for SE (Spin-Echo signal sequence) is shown in follow equation (1) [2]:

$$S = M_0 * (1 - \exp(\frac{-TR}{T1})) * (\exp(\frac{-TE}{T2})) \tag{1}$$

where: M_0 - initial magnetization or start state, TR – Time to Repeat, TE – Echo Time, T1 – relaxation time, T2 – relaxation time, S – MRI signal.

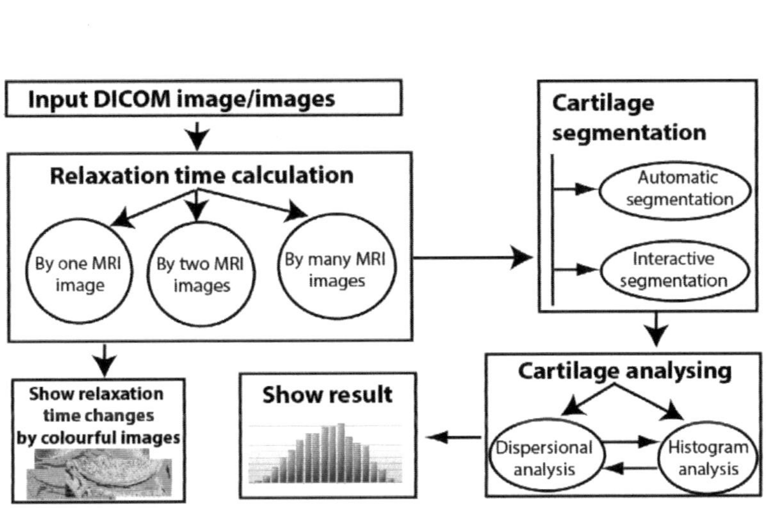

Figure 2. Common flowchart.

The initial magnetization(M_0) is foremost depending on proton density and magnetic field power (B_0) measured in tesla(T). This is shown in equation (2). Usually is using MRI with 1 T, 1.5 T and 3 T.

$$M_0 \sim \rho * B_0 \tag{2}$$

where: ρ - proton density, B_0 - magnetic field power.

Manipulate with TR and TE parameters we can change relaxation times and proton density influence on MRI signal. In the manipulation result with these parameters we can get three types MRI images: T1 weighted, T2 weighted and proton density (PD) image (table 1) [1]. According to MRI image type PD/T1/T2 the MRI signal foremost is depending on proton density, relaxation time(T1) relaxation time(T2) [11].

Table 1.TR and TE parameters for SE sequence.

Sequence	MRI image type	Parameters TR	Parameters TE
SE	T1	600 ms	10-30 ms
	PD	1000 ms	10-30 ms
	T2	2000 ms	80-250 ms

Usually a doctor gets from MRI DICOM images. Each DICOM image pixels contain intense value which is proportional to MRI signal (S). Assume that MRI signal is approximately equal with DICOM image pixel containing intense value (SI). This is shown in follow equal: $SI \approx S$.

When we get MRI images we know only MRI images pixels values (SI) and images parameters TE and TR. But we want to know T2 and T1 relaxation times.

3.1. Relaxation time calculation by one MRI image

This method "relaxation time calculation by one MRI image" is very useful for doctors, because of doctors that usually use only one MRI image which shows only one slice. This method is very efficient because the time it takes to get an MRI image decreases. One MRI image is easier to obtain than multiple ones. If we use this method, we'll be able to save MRI working time and reduce RF (radio frequency) influence on patient health.

But this method has some disadvantages, as it doesn't bear in mind some parameters which influence on pixels intense value (SI). As a result T1 and T2 relaxation time calculation is very approximate. That method is useful for T1 and T2 changes displaying.

For T2 and T1 times calculation is assumed that proton density (ρ) is constant and magnetic field power (B_0) is constant. If assumed that density (ρ) is constant and magnetic field power (B_0) is constant, than initial magnetization (M_0) is constant.

If we want calculate T2 relaxation time, than we must use T2 weighted MRI image. When we use T2 weighted MRI image we can don't consider the second part of equation (1): (1 – exp(-TR/T1)). So, now we have the follow equation (3):

$$T2 = \frac{-TE}{\ln\left(\frac{SI}{SI_{MAX}}\right)} \tag{3}$$

where $SI_{MAX} \approx M_0$.

Now we can calculate T2. For T1 calculation we can use the same method.

3.2. Relaxation time calculation by two MRI images

For relaxation time calculation we can use two MRI images [4]. When we have use two MRI images, than we have more information and we can calculate T1/T2 relaxation time more precise than in the previous method. For T1 calculating it possible use the follow equation (4):

$$T1 = \frac{-TR_1}{\ln\left(\frac{SI_2 - SI_1}{SI_1}\right)} \tag{4}$$

where: TR_1 - Time to Repeat of first image, SI_1 - intense value of first image, SI_2 – intense value of second image, $TR_2 = 2*TR_1$ and $TE_1 = TE_2$. There is a proof (5) of equation (4): $-TR_1 / T1$ are denoted by letter (a)

$$\frac{SI_2 - SI_1}{SI_1} = \frac{M_0 * (1-\exp(2a)) * \exp(\frac{-TE}{T2}) - M_0 * (1-\exp(a)) * \exp(\frac{-TE}{T2})}{M_0 * (1-\exp(a)) * \exp(\frac{-TE}{T2})} =$$
$$= \frac{(1-\exp(2a)) - (1-\exp(a))}{(1-\exp(a))} = \frac{-\exp(a)*\exp(a)+\exp(a)}{(1-\exp(a))} = \frac{\exp(a)*(1-\exp(a))}{(1-\exp(a))} = \exp(\frac{-TR_1}{T1}) \tag{5}$$

3.3. Relaxation time calculation by many MRI images

For relaxation time calculation we can use many MRI images [3]. This method gets possibility modulate relaxation process. Usually when we use this method we have 7 or 8 images, which show one slice. These images have different TE parameters (when we want calculate T2), and TR parameters (when we want calculate T1). So, for each MRI images pixel we have 7 or 8 intense values, which show signal intense in different times. This gives us the possibility approximate relaxation process. The approximation we can make for two parameters: initial magnetization (M0) and relaxation time (T2). For approximation task solving can use least square method and partial derivatives. This we can see from the following formula (6):

$$S(P1,P2)=\sum_{i=1}^{n}(SI_i - P1*\exp(\frac{-TE_i}{P2}))^2 \rightarrow \min$$

$$\begin{cases} \dfrac{\partial S}{\partial P1} = -2\sum_{i=1}^{n}(e^{\frac{-TE_i}{P2}} *(SI_i - P1*e^{\frac{-TE_i}{P2}})) = 0 \\ \dfrac{\partial S}{\partial P2} = -\dfrac{2}{P2^2}*P1*\sum_{i=1}^{n}(e^{\frac{-TE_i}{P2}}*(SI_i - P1*e^{\frac{-TE_i}{P2}})*TE_i) = 0 \end{cases} \quad (6)$$

where: P1 - initial magnetization (**M_0**), P2 - relaxation time (**T2**).

3.4. Cartilage segmentation

The cartilage segmentation is useful procedure for cartilage analyzing. The cartilage segmentation aim is detect cartilage pixels. When we work with MRI images we have a lot of segmentation problems: different images planes, many MRI parameters, which strongly change MRI signal, different sequence, artifacts, patient movement and noise. Therefore, in the experiments has been used interactive segmentation method, which is universal and give possibility use different MRI images.

Interactive method can make different organic structure segmentation. For this method developing was taken two other methods clusterization segmentation and region growing segmentation. These methods symbiosis give very good result.

3.5. Cartilage analysing

After successful cartilage segmentation is possible make cartilage analyzing. For cartilage analyzing we can use dispersional analysis and histogram analysis.

4. Experimental Results

In this experimental part is shown that proposed method allow detect OA grade and find cartilage pathogen zone.

A doctor can't see the first grade of the OA due to he use simple grayscale MRI images. But the knee articular cartilage changes are recoverable at the first grade of the OA. So, OA first grade displaying is very important. In figure 5 is shown that proposed methods allow calculate and displaying relaxation times changes which is marked by red color. These results show that thanks to proposed methods a doctor can see cartilage pathogen zone and make OA early diagnostic.

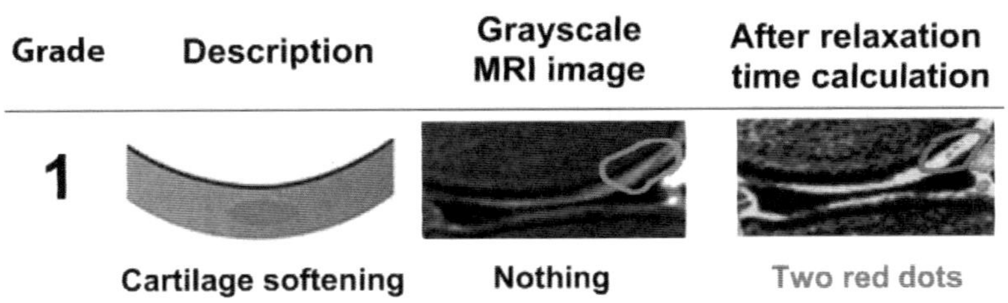

Figure 5. OA early stage detection.

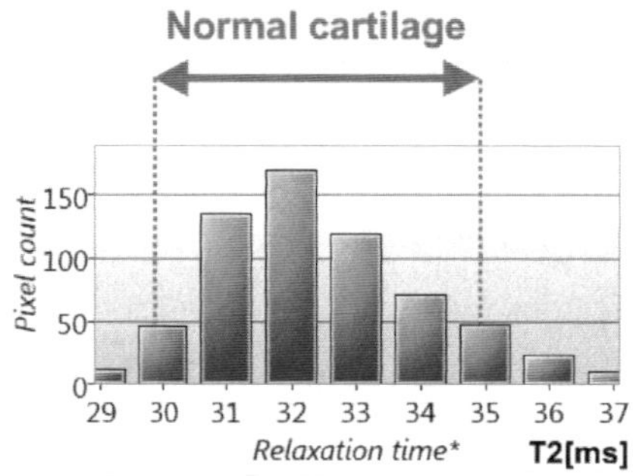

Figure 6. Healthy cartilage.

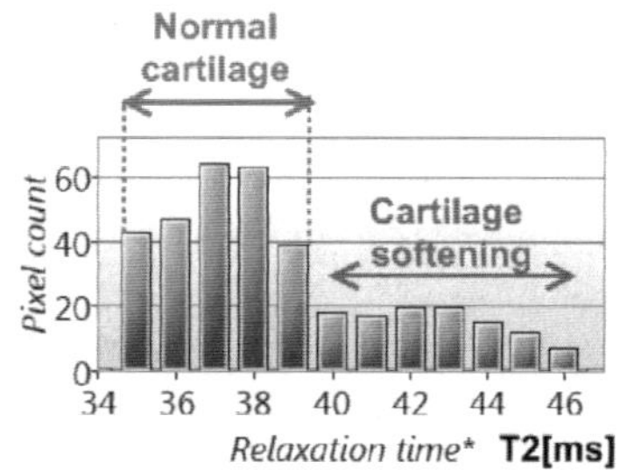

Figure 7. OA first grade.

In the figures 6 is shown healthy histogram cartilage. In the figures 7 is shown fist grade of OA. Each histogram has unique form. Therefore, the OA grades can be detected by histograms unique forms. Each OA grades have unique histograms forms.

The experimental results is shown that patient with damaged cartilage has grater disperse than patient with healthy cartilage. In figure 8 we can see six patients. The sixth patient hasn't problems, because of this patient disperse is the smallest.

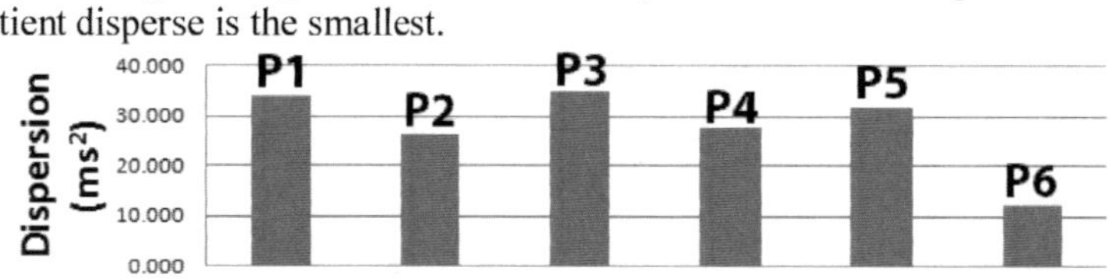

Figure 8. The healthy patient has smallest dispersion. P1...P6 - patient number.

5. Conclusion

In this paper, methods are proposed, which allow to make an early osteoarthritis diagnostic. Proposed methods consist of three parts: relaxation time calculation, cartilage segmentation, cartilage analysing.

The most commonly is used many MRI images (7 or 8 images) for relaxation time calculation [2, 14]. Proposed relaxation time calculations methods are universal, that allow calculating relaxation time by any MRI images count. It is very useful for old MRI, which can't get many MRI images in the same moment. "Relaxation time calculation by one MRI image" method is very effective, because the time of MRI imaging is decreasing. One MRI image we can get faster than many images. If we use this method, will be able to save MRI working time and reduce RF (radio frequency) influence on patient health. When we have use more MRI images, than we have more information and we can calculate T1/T2 relaxation time more precise. "Relaxation time calculation by many MRI images" method gets possibilities modulate relaxation process. All these methods allow calculate T2/T1 relaxation times. After it is important to show relaxation time changes by colorful images, that allow displaying relaxation times changes. Experimental results show that thanks to proposed methods a doctor can see cartilage pathogen zone and make OA early diagnostic.

References

[1] Evert B 2004 *Basic MRI Physics* p 76
[2] Hornak J P 1997 *The Basics of NMR*
[3] Bruce C Po 2001 *Graphical Interface for Quantitative T1 and T2 Mapping of MRI Data* (Massachusetts Institute of Technology) p 119
[4] Fanea L and Sfrangeu S A 2010 *Relaxation times mapping using magnetic resonance imagin*
[5] Murphy L and Helmick C G 2012 The impact of osteoarthritis in the United States: a population-health perspective *American Journal of Nursing* **112** S13-S19

[6] Mallikarjuna M S and Mallikarjun S 2012 Knee joint articular cartilage segmentation, visualization and quantification using image processing techniques: a review *International Journal of Computer Applications* **42** 36-43
[7] Dam E B, Folkesson J, Pettersen P C and other 2006 *Semi-Automatic Knee Cartilage Segmentation* p 9
[8] Fripp J, Lucas E and Engstrom C 2013 *Automated Morphological Knee Cartilage Analysis of 3D MRI at 3T*
[9] Tamez-Pena J G, Barbu-McInnis M and Totterman S 2004 *Knee Cartilage Extraction and Bone-Cartilage Interface Analysis from 3D MRI Data Sets*
[10] Nasnikova I U, Morozov S P and Filisteev P A 2011 Magnetic resonance tomography: quantitative assessment methods in articular cartilage structure analysis of in patients with osteoartrosis *Russian Electronic Journal of Radiology* **1** 75-81
[11] Haacke E M, Brown R W, Thompson M R and Venkatesan R 1999 *Magnetic Resonance Imaging Physical Principles and Sequence Design* (New York: Wiley-Liss) p 973
[12] Roach H I and Tilley S 2007 *The Pathogenesis of Osteoarthritis*
[13] Klussmann A, Gebhardt H, Liebers F, Lars V E and others 2008 *Individual and Occupational Risk Factors for Knee Osteoarthritis – Study Protocol of a Case Control Study*
[14] Blumenkrantz G and Majumdar S 2007 Quantitative magnetic resonance imaging of articular cartilage in osteoarthritis *European Cell and Materials* **13** 75-86

Impulsive Vaccination for an Epidemiology Model

M. De la Sen[1], **A. J. Garrido**
Institute of Research and Development of Processes,
University of the Basque Country,
Bilbao, Spain

E-mail: manuel.delasen@ehu.es

Abstract. This paper investigates sufficient conditions of almost periodic sand periodic solutions of an integral model under impulsive controls. Since the model is of generic epidemiological interest, such impulsive controls are either vaccination actions or abrupt variations of the infected population due to infected immigration or lost of infective numbers due to either vaccination or lost of infected population by out-migration.

1. Introduction

This research deals with a nonlinear integral equation with eventual distributed time-varying time delay. Some of its necessary and relevant properties like positivity and boundedness of the solution are investigated. Also, some conditions of almost periodicity and periodicity of the solutions are dealt with under the action of impulsive controls art certain impulsive time instants. The impulsive controls can be interpreted as vaccination effort or infected immigration if the infected population decreases or output migration otherwise. The model has been investigated in [1-9] in a simplified version the impulsive- free case. The impulsive vaccination is a control, although the epidemic model is not controllable, by nature. For state and output measuring-based controller synthesis, see, for instance, [10-13]. The following basic notation is used:

$\boldsymbol{R}_+ = \{ z \in \boldsymbol{R} : z > 0 \}$, $\boldsymbol{R}_- = \{ z \in \boldsymbol{R} : z < 0 \}$, $\boldsymbol{R}_{0+} = \boldsymbol{R}_+ \cup \{0\}$, $\boldsymbol{R}_{0-} = \boldsymbol{R}_- \cup \{0\}$, $\boldsymbol{R}_{0+}^n$ and $\boldsymbol{R}_+^n$ are the n-th orthants of $\boldsymbol{R}^n$ whose components are in $\boldsymbol{R}_{0+}$ and $\boldsymbol{R}_+$, respectively. $\boldsymbol{R}_{0+}^{n\times n}$ and $\boldsymbol{R}_+^{n\times n}$ denote the sets of real square n-matrices whose entries are in $\boldsymbol{R}_{0+}$ and $\boldsymbol{R}_+$, respectively. $\boldsymbol{R}_{0-}^n$ and $\boldsymbol{R}_-^n$ are the n-th orthants of $\boldsymbol{R}^n$ whose components are in $\boldsymbol{R}_{0-}$ and $\boldsymbol{R}_-$, respectively. $\boldsymbol{R}_{0-}^{n\times n}$ and $\boldsymbol{R}_-^{n\times n}$ denote the sets of real square n-matrices whose entries are in $\boldsymbol{R}_{0-}$ and $\boldsymbol{R}_-$, respectively,

$\boldsymbol{Z}_+ = \{ z \in \boldsymbol{Z} : z > 0 \}$, $\boldsymbol{Z}_{0+} = \{ z \in \boldsymbol{Z} : z \geq 0 \} = \boldsymbol{Z}_+ \cup \{0\}$, $\overline{n} = \{1, 2, \dots, n\}$,

if $x, y \in \boldsymbol{R}^n$ then $x \leq y$ $(x < y)$ means $x_i \leq y_i$ $(x_i < y_i)$; $\forall i \in \overline{n}$. If $A, B \in \boldsymbol{R}^{n\times m}$ then $A = (A_{ij}) \leq B = (B_{ij})$ $(A < B)$ means $A_{ij} \leq B_{ij} (A_{ij} < B_{ij})$; $\forall (i, j) \in \overline{n} \times \overline{m}$,

the superscript T stands for the transpose of a real vector or matrix,

---

[1] To whom any correspondence should be addressed. This research is supported by the Spanish Government through Grants DPI2012-30651 and DPI2015-64766-R

if $x \in \boldsymbol{R}^n$, then $|x| = (|x_1|, |x_2|, \dots, |x_n|)^T$, if $A \in \boldsymbol{R}^{n \times m}$ then $|A| = (|A|_{ij}) = (|A_{ij}|)$, if a real function $x(t)$ is discontinuous at some t then $x(t^-)$ is its left limit at t and $x(t)$ (abbreviating $x(t^+)$) is its right limit so that if it is continuous at t its value is $x(t) = x(t^-)$.

2. The model and its impulsive controlling actions

Consider the nonlinear integral equation with a finite distributed delay function $\tau : \boldsymbol{R}_{0+} \to \boldsymbol{R}_+$ and eventual bounded discontinuities:

$$x(t^-) = \int_{t-\tau(t)}^{t} F(s, x(s), y(s))ds \ ; \quad x(t) = K(t)x(t^-) \tag{1}$$

, [1-2], where $\tau(0) \ge \sup_{t \in \boldsymbol{R}_{0+}} \tau(t)$, $F : (\boldsymbol{R}_{0+} \cup [-\tau(0), 0)) \times \boldsymbol{R}_{0+}^n \times \boldsymbol{R}_{0+}^n \to \boldsymbol{R}_{0+}^n$, $x, y : [-\tau(0), \infty) \to \boldsymbol{R}^n$, and $K : \boldsymbol{R}_{0+} \to \boldsymbol{R}_{0+}^{n \times n}$ are a positive vector function and a positive square n-matrix function, respectively. In particular, $x : [-\tau(0), \infty) \to \boldsymbol{R}^n$ is the solution of (1) and $y : [-\tau(0), \infty) \to \boldsymbol{R}^n$ can describe different actions like, for instance, an external disturbance (i.e. a disturbance which is independent of the solution) or either an internal solution-dependent disturbance or a point-delay or distributed delayed or nonlinear contribution to the solution. For instance, the special form $F(t, x, y) = f(t, x) + g(t, y)$ is discussed in [2] for the integral equation (1), where f, g and τ are positive almost periodic. Integral equations of the form (1) are of interest in the descriptions of some epidemic models. For instance, $x(.)$could be the susceptible population which is measurable or partially known and can be treated under vaccination effort, and $y(.)$ some coupled population like, for instance, the infectious one. Note that, for any $t \in \boldsymbol{R}_{0+}$, if $K(t) = I_n$ (the n-th identity matrix) then $x(t) = x(t^-)$ so that $x : [-\tau(0), \infty) \to \boldsymbol{R}^n$ is continuous at time t. If $K(t) \ne I_n$ then $x(t) \ne x(t^-)$ if $x(t^-) \ne 0$. Note that (2) versus (1) can be equivalently viewed for the case when $x \equiv y$ under an integral contribution of an impulsive control gain $K(t)$ as

$$x(t) = x(t^-) + \int_{t^-}^{t} (K(s) - I_n)x(s)\delta(s-t)ds = x(t^-) + \lim_{\varepsilon \to 0^+} \int_{t-\varepsilon}^{t+\varepsilon} (K(s) - I_n)x(s)\delta(s-t)ds = x(t^-) + (K(t) - I_n)x(t^-) \tag{2}$$

where the integrals in (1)-(2) are Lebesgue integrals, $(K(t) - I_n)x(t^-)$ is the left-continuous test vector function which has a compact support and $\delta(t)$ denotes the Dirac distribution such that

$\int_{-\infty}^{\infty} \delta(s)ds = \lim_{\varepsilon \to 0^+} \int_{-\varepsilon}^{\varepsilon} \delta(s)ds = 1$.Then $\delta(t) = +\infty$ if $t = 0$ and $\delta(t) = 0$ if $t \ne 0$. Note that denoting the shifted Dirac delta as $\delta_t(s) = \delta(s-t)$, one has for any $t \in \boldsymbol{R}_{0+}$

$$\int_{-\infty}^{\infty} f(s)\delta(s-t)ds = \lim_{\varepsilon \to 0^+} \int_{t-\varepsilon}^{t+\varepsilon} f(s)\delta(s-t)ds = \int_{t^-}^{t} f(s)\delta(s-t)ds = \int_{t^-}^{t} f(s)\delta_t(s)ds = f(t^-)$$

Note also that

$\int_{-\infty}^{\infty} f(s)\delta(s)ds = \lim_{\varepsilon \to 0^+} \int_{-\varepsilon}^{\varepsilon} f(s)\delta(s-t)ds = f(0^-)$ so that the Dirac distribution delta $\delta(t)$ is also a not absolutely continuous measure with argument being a subset A of the real line with $\delta(A) = 1$ if $0 \in A$ and $\delta(A) = 1$ if $0 \notin A$. The impulsive set of time instants is $Imp = \{t \in \boldsymbol{R}_{0+} : x(t) \ne x(t^-) \ \ if \ \ x(t^-) \ne 0\}$.

The following positivity result for the solution (1)-(2), in the sense that no component state takes negative values at any time if the initial conditions are non-negative, is immediate:

Proposition 1. Assume some initial condition $x:[-\tau(0),0)\to \boldsymbol{R}_{0+}^{n}$ for the model with $F:(\boldsymbol{R}_{0+}\cup[-\tau(0),0))\times \boldsymbol{R}_{0+}^{n}\times \boldsymbol{R}_{0+}^{n}\to \boldsymbol{R}_{0+}^{n}$ in (1) such that the feedback control gain $K:\boldsymbol{R}_{0+}\to \boldsymbol{R}_{0+}^{n\times n}$. Then, the following properties hold:

(i) $x:[-\tau(0),\infty)\to \boldsymbol{R}_{0+}^{n}$,

(ii) if the feedback control is replaced with the non-feedback control $x(t)=K(t)$ if $t\in Imp$ with $K:\boldsymbol{R}_{0+}\to \boldsymbol{R}_{0+}^{n}$ then $x:[-\tau(0),\infty)\to \boldsymbol{R}_{0+}^{n}$.

Note that the impulsive controlled action can be interpreted as due to a culling or quarantine of the infected population if $\|K(t)\|<1$. Contrarily, if $K(t)$ causes the inequality $\|x(t)\|>\|x(t^-)\|$ then there is a positive increase of infected population due to infected immigration into the studied environment.

3. Periodic and almost periodic solutions

Now, consider the impulsive set of time instants t_i, $t_{i+1}(>t_i)\in Imp\subset \boldsymbol{R}_{0+}$, with $T_i=t_{i+1}-t_i$, so that for some sequence $\{K(t_i)\}$ of non-identity matrices, one has:

$$x(t_i)=K(t_i)x(t_i^-)=\int_{t_i-\tau(t_i)}^{t_i}F(s,x(s),y(s))ds\ ;\ \forall t_i\in Imp \tag{3}$$

The impulsive set Imp is denumerable and consists of strictly ordered elements and can be finite or infinite. Introduce by convenience the identities:

$$\tau(t_{i+1})=\tau_0+\tilde{\tau}(t_{i+1})=\tau_0+\tilde{\tau}(t_i)+\tilde{\tilde{\tau}}(t_i);\quad K(t_{i+1})=K_0+\tilde{K}(t_{i+1})=K_0+\tilde{K}(t_i)+\tilde{\tilde{K}}(t_i)=K(t_i)+\tilde{\tilde{K}}(t_i) \tag{4}$$

; $\forall t_i\in Imp$, which allow to refer all delays $\tau(t_{i+1})$ at the impulsive set related to a constant delay τ_0 plus its deviation $\tilde{\tau}(t_{i+1})$. It is assumed that the nominal gain $K:\boldsymbol{R}_{0+}\to \boldsymbol{R}_{0+}^{n\times n}$ This deviation can be related to that of its delay $\tilde{\tau}(t_i)$ at the preceding impulsive time instant by its incremental value $\tilde{\tilde{\tau}}(t_i)=\tilde{\tau}(t_{i+1})-\tilde{\tau}(t_i)$. A similar incremental description can be given for the impulsive gain sequence. The following definitions are then used to characterize almost periodicity of the solutions, [2].

Definition 1. A set $E\subset \boldsymbol{R}$ of real numbers is said to be relatively dense if there exists a number $\ell\in \boldsymbol{R}_+$ such that any real interval $(\alpha,\alpha+\ell)$ of length ℓ contains at least one number in E.

Definition 2. A function $f\in BC(\boldsymbol{R}\times\boldsymbol{\Omega},\boldsymbol{R}^q)$ is said to be almost periodic in $t\in\boldsymbol{R}$ uniformly in $x\in\boldsymbol{\Omega}$ (abbreviated by $f\in AP(\boldsymbol{R}\times\boldsymbol{\Omega},\boldsymbol{R}^q)$) if the ε-translation set of f

$$T(f,\varepsilon)=\{\omega\in\boldsymbol{R}:\|f(t+\omega,z)-f(t,z)\|<\varepsilon;\forall(t,z)\in\boldsymbol{R}\times\boldsymbol{W},\forall compact\ set\ \boldsymbol{W}\subset\boldsymbol{\Omega}\}$$

is a relatively dense set in $\boldsymbol{R}$ for all $\varepsilon\in\boldsymbol{R}_+$. Each $\omega\in T(f,\varepsilon)$ is called an ε- period of f.

Almost periodic functions without uniformity in $\boldsymbol{\Omega}$ are defined closely by removing the argument z in $\boldsymbol{\Omega}$ and the compact subset $\boldsymbol{W}$ from the definition of $T(f,\varepsilon)$. See, for instance, [2].

Definition 3. A function $f\in BC(\boldsymbol{R}\times\boldsymbol{\Omega},\boldsymbol{R}^q)$ is said to be ρ-almost periodic in $t\in\boldsymbol{R}$ uniformly in $x\in\boldsymbol{\Omega}$ (abbreviated by $f\in AP(\mathbf{R}\times\mathbf{\Omega},\mathbf{R}^q,\rho)$) for some $\rho\in\mathbf{R}_+$ if the ε-translation set of f

$T(f,\varepsilon)=\{\omega\in \boldsymbol{R}: \| f(t+\omega, z)-f(t, z)\| <\varepsilon; \forall(t,z)\in \boldsymbol{R}\times \boldsymbol{W}, \forall compact\ set\ \boldsymbol{W}\subset \boldsymbol{\Omega}\}$
is a relatively dense set in $\boldsymbol{R}$ for all $\varepsilon\in[\rho,\infty)$. Each $\omega\in T(f,\varepsilon)$ is called an ε- period of f for a given $\varepsilon\in[\rho,\infty)$.

The following main result, whose proof is omitted, follows:

Theorem 1. Assume that the model integral equation, subject to $F:(\boldsymbol{R}_{0+}\cup[-\tau(0),0))\times \boldsymbol{R}_{0+}^n\times \boldsymbol{R}_{0+}^n\to \boldsymbol{R}_{0+}^n$, $x,y:[-\tau(0),\infty)\to \boldsymbol{R}^n$ and $K:\boldsymbol{R}_{0+}\to \boldsymbol{R}_{0+}^{n\times n}$, satisfies the subsequent hypotheses:

H1) $F\in BC(\boldsymbol{R}_{0+}\times \boldsymbol{R}_{0+}^n, \boldsymbol{R}_{0+}^n)\cap AP(\boldsymbol{R}_{0+}\times \boldsymbol{R}_{0+}^n, \boldsymbol{R}_{0+}^n,\varepsilon)$ for some given $\varepsilon\in \boldsymbol{R}_+$,

H2) $T_i=t_{i+1}-t_i\ge \tau_0+\tilde{\tau}(t_i)$, the i- th inter-impulse time interval $T_i\in P(F,\ \lambda_{1i}\varepsilon)$ (i.e. it is an $\lambda_{1i}\varepsilon$- period for F for some given $\varepsilon\in \boldsymbol{R}_+$), $(T_i-\tau_0)\in P(F,\ \lambda_{2i}\varepsilon)$, $(T_i-\tau_0-\tilde{\tau}(t_i))\in P(F,\ \lambda_{3i}\varepsilon)$ for some $\lambda_{ji}\in \boldsymbol{R}_+$ and $j\in\bar{3}$; $\forall i\in \boldsymbol{Z}_{0+}$, for $\varepsilon\in \boldsymbol{R}_+$ in H1,

H3) $\tau_0\lambda_{1i}+\tilde{\tau}(t_i)\lambda_{2i}+\tilde{\tilde{\tau}}(t_i)\lambda_{3i}\le 1$; $\forall i\in \boldsymbol{Z}_{0+}$

Thus, the following properties hold: **(i)** $\|\tilde{x}(t_i^-)\|<\varepsilon$ and $\{T_i\}$ is a discrete ε-period sequence for the left limit solution $x(t^-)$. If $\|K(t_i)\|\le min\left(\frac{1}{2}, \frac{1-\|K(t_{i-1})\|}{\|x(t_i^-)\|}\varepsilon-\|K(t_{i-1})\|\right)$ then $\|\tilde{x}(t_i)\|<\varepsilon$ and $\{T_i\}$ is a discrete ε-period sequence for the right limit solution $x(t)$,

(ii) $\|\tilde{x}(t_{i\eta})\|<\varepsilon$, where $\tilde{x}(t_{i\eta})=x(t_i+\eta)-x(t_i)$, for any real $\eta\in(0,T_i)$; $\forall i\in \mathbf{Z}_{0+}$ so that $x\in AP(\boldsymbol{R}\times \boldsymbol{R}_{0+}^n, \boldsymbol{R}_{0+}^n,\varepsilon)$, i.e. the solution of (1)-(2) is ε-almost periodic in $t\in \boldsymbol{R}$ uniformly in $\boldsymbol{R}_{0+}^n$.

Under the conditions of Theorem 1, the free-impulsive integrand of the model is assumed almost-periodic according to the Hypothesis H1. However, the impulsive effort at certain impulsive time instants can also achieve the almost periodicity of the solution. The following relations are obtained:

$$\begin{aligned}
x(t+\rho)-x(t) &= \textstyle\int_{t+\rho-\tau(t)}^{t+\rho} F(s,\mu(s)x_{if}(s),V(s,x(s)))ds-\int_{t-\tau(t)}^{t} F(s,\mu(s)x_{if}(s),V(s,x(s)))ds\\
&\ge \lambda\left(x_{if}(t+\rho)-x_{if}(t)\right)=\lambda\left(x_{if}(t+\rho)-x_{if}(t)\right)+\lambda_0(t,\rho)\\
&+\textstyle\sum_{i=1}^{m(t+\rho)}F(t_i(t+\rho),x(t_i^-(t+\rho)),0)V(t_i(t+\rho),x(t_i^-(t+\rho)))-\sum_{i=1}^{m(t)}F(t_i(t),x(t_i^-(t)),0)V(t_i(t),x(t_i^-(t)))\\
&\le \lambda\,\varepsilon_{if}+\lambda_0(t,\rho)-\textstyle\sum_{i=1}^{m(t+\rho)}F(t_i(t+\rho),x(t_i^-(t+\rho)),0)\left|V(t_i(t+\rho),x(t_i^-(t+\rho)))\right|\\
&+\textstyle\sum_{i=1}^{m(t)}F(t_i(t),x(t_i^-(t)),0)\left|V(t_i(t),x(t_i^-(t)))\right|\le\varepsilon
\end{aligned}\tag{5}$$

provided that under impulsive vaccination (i.e. negative impulses) :

$$\textstyle\sum_{i=1}^{m(t+\rho)}F(t_i(t+\rho),x(t_i^-(t+\rho)),0)\left|V(t_i(t+\rho),x(t_i^-(t+\rho)))\right|$$

$$-\sum_{i=1}^{m(t)} F\left(t_i(t), x\left(t_i^-(t)\right), 0\right)\left|V\left(t_i(t), x\left(t_i^-(t)\right)\right)\right| \geq \lambda\, \varepsilon_{if} + \lambda_0(t, \rho) - \varepsilon \tag{6}$$

and under general impulsive action with non-necessarily negative impulses:

$$\begin{aligned}&\sum_{i=1}^{m(t+\rho)} F\left(t_i(t+\rho), x\left(t_i^-(t+\rho)\right), 0\right) V\left(t_i(t+\rho), x\left(t_i^-(t+\rho)\right)\right)\\ &\quad -\sum_{i=1}^{m(t)} F\left(t_i(t), x\left(t_i^-(t)\right), 0\right) V\left(t_i(t), x\left(t_i^-(t)\right)\right) \leq \varepsilon - \lambda\varepsilon_{if} - \lambda_0(t, \rho)\end{aligned} \tag{7}$$

Let the integrand of the epidemic model be decomposable as $F(t, x, y) = f(t, x) + g(t, y)$ subject to

$$g\left(t, \alpha^{-1}x\right) \geq \varphi(t, \alpha, x) g(t, x) > \alpha\, g(t, x) \geq \beta g(t, x)$$
$$f(t, \beta x) \geq \varphi\left(t, \beta^{-1}, x\right) f(t, x) > \beta f(t, x)$$

with $min\left(\varphi(t, \alpha, x), \varphi\left(t, \alpha^{-1}, x\right)\right) > \alpha \geq \beta$ and $min\left(\varphi(t, \beta, x), \varphi\left(t, \beta^{-1}, x\right)\right) > \beta$. Then,

$$min\left(\varphi(t, \alpha, x), \varphi\left(t, \beta^{-1}, y\right)\right) > \beta$$

so that

$$F\left(t, \beta x, \alpha^{-1}y\right) = f(t, \beta x) + g\left(t, \alpha^{-1}y\right) > \beta\left(f(t, x) + g(t, y)\right) = \beta F(t, x, y)$$

where $\beta, \alpha \in (0, 1)$ with $\beta \leq \alpha$. Assume that the scalar vaccination gains are generated as follows:

$$k_{di}(t) = -\left|k_{di}(t)\right| sgn\, F_i(t, s, x(s), x(s-\theta))$$

then,

$$\begin{aligned}&x\left(t^-\right) = \int_{t-\tau(t)}^{t^-} F(t, s, x(s), x(s-\theta))\, ds\\ &x_i(t) = x_i\left(t^-\right) - \int_{t^-}^{t} \left|F_i(t, s, x(s), x(s-\theta(s)))\right| \left|k_i(t)\right| \delta(t-s)\, ds\\ &= x_i\left(t^-\right) - \left|k_i(t)\right| \left|F_i\left(t, t, x\left(t^-\right), x(t-\theta(t))^-\right)\right|\end{aligned}$$

If $x\left(t^-\right) = \alpha^{-1}(t) x(t)$ and $\beta(t)$ are the impulsive infective population increase at time $(t-\theta(t))$ and the corresponding impulsive vaccination correction at the current time t satisfying $\beta(t) \leq \alpha(t-\theta(t)) < 1$

$$x\left(t^-\right) - x(t) = |k(t)| \left|F\left(t, t, x\left(t^-\right), x(t-\theta(t))^-\right)\right| \geq |k(t)| \beta(t) \left|F(t, t, x(t), x(t-\theta(t)))\right| \tag{8}$$

so that

$$\begin{aligned}x(t+\rho) &= \int_{-\tau_0-\tilde{\tau}(t+\rho)}^{0} F\left(s+t+\rho, x(s+t+\rho), V\left(s+t+\rho, x\left(s^-+t+\rho\right)\right)\right) ds\\ &= \int_{-\tau(t+\rho)}^{0} F(s+t+\rho, x(s+t+\rho), 0)\, ds\\ &\quad + \sum_{t_i(t+\rho) \in Imp[t+\rho-\tau_0-\tilde{\tau}(t+\rho), t+\rho]} F\left(t_i(t+\rho), x\left(t_i^-(t+\rho)\right), 0\right) V\left(t_i(t+\rho), x\left(t_i^-(t+\rho)\right)\right)\end{aligned} \tag{9}$$

for $t \geq \tau_0 - \tau(0) + \tilde{\tau}_M$ with $\tau(t) = \tau_0 + \tilde{\tau}(t)$, $\tilde{\tau}(t) \in [\tilde{\tau}_m , \tilde{\tau}_M]$, $\tilde{\tau}_M \geq \tilde{\tau}_m \geq -\tau_0$ and $\rho \in \boldsymbol{R}_{0+}$ .Note that

$$\left\| x(t) - x(t^-) \right\| = \begin{cases} 0 & if \quad t \notin Imp[t - \tau(t), t] \\ \left\| F(t, x(t^-), 0) V(t, x(t^-)) \right\| & if \quad t \in Imp[t - \tau(t), t] \end{cases}$$

$$\left\| x(t^- + \rho) - x(t) \right\| = \left\| x(t^- + \rho) - x(t^-) - (x(t) - x(t^-)) \right\| \leq \begin{cases} \left\| x(t^- + \rho) - x(t) \right\| & if \quad t \notin Imp[t - \tau(t), t] \\ \left\| x(t^- + \rho) - x(t) \right\| + \left\| F(t, x(t^-), 0) V(t, x(t^-)) \right\| & if \quad t \in Imp[t - \tau(t), t] \end{cases} \tag{10}$$

On the other hand, if $\tau(t+\rho) = \tau(t)$, $F(t, x(t), 0)$ is Lipschitz with constant K

$$\begin{aligned} \left\| x(t^- + \rho) - x(t^-) \right\| \leq & K \int_{-\tau(t)}^{0} \left\| x(s + t + \rho) - x(s + t) \right\| ds \\ & + \left\| \sum_{t_i(t+\rho) \in Imp[t+\rho-\tau(t), t+\rho)} F(t_i(t+\rho), x(t_i^-(t+\rho)), 0) V(t_i(t+\rho), x(t_i^-(t+\rho))) \right. \\ & \left. - \sum_{t_i(t) \in Imp[t-\tau(t), t)} F(t_i(t), x(t_i^-(t)), 0) V(t_i(t), x(t_i^-(t))) \right\| \end{aligned} \tag{11}$$

$$\begin{aligned} \left\| x(t^- + \rho) - x(t) \right\| \leq & K \int_{-\tau(t)}^{0} \left\| x(s + t + \rho) - x(s + t) \right\| ds \\ & + \left\| \sum_{t_i(t+\rho) \in Imp[t+\rho-\tau(t), t+\rho)} F(t_i(t+\rho), x(t_i^-(t+\rho)), 0) V(t_i(t+\rho), x(t_i^-(t+\rho))) \right. \\ & \left. - \sum_{t_i(t) \in Imp[t-\tau(t), t]} F(t_i(t), x(t_i^-(t)), 0) V(t_i(t), x(t_i^-(t))) \right\| \end{aligned} \tag{12}$$

$$\left\| x(t + \rho) - x(t) \right\| \leq \begin{cases} \left\| x(t^- + \rho) - x(t) \right\| & if \quad t + \rho \notin Imp[t + \rho - \tau(t + \rho), t + \rho] \\ \left\| x(t^- + \rho) - x(t) \right\| + \left\| F(t + \rho, x(t^- + \rho), 0) V(t + \rho, x(t^- + \rho)) \right\| & if \quad t + \rho \in Imp[t + \rho - \tau(t + \rho), t + \rho] \end{cases} \tag{13}$$

$$\begin{aligned} \left\| x(t + \rho) - x(t) \right\| \leq & K \int_{-\tau(t)}^{0} \left\| x(s + t + \rho) - x(s + t) \right\| ds \\ & + \left\| \sum_{t_i(t+\rho) \in Imp[t+\rho-\tau(t), t+\rho]} F(t_i(t+\rho), x(t_i^-(t+\rho)), 0) V(t_i(t+\rho), x(t_i^-(t+\rho))) \right. \\ & \left. - \sum_{t_i(t) \in Imp[t-\tau(t), t]} F(t_i(t), x(t_i^-(t)), 0) V(t_i(t), x(t_i^-(t))) \right\| \end{aligned} \tag{14}$$

$$I_{imp}(t, x_t) = \sum_{t_i(t) \in Imp[t-\tau(t), t]} V(t_i(t), x(t_i^-(t))) \delta(t - t_i(t)) \tag{15}$$

Thus, the following result holds:

Theorem 2. Assume that $\tau(t+\rho) = \tau(t)$, $F(t, x(t), 0)$ is Lipschitzian with constant K. Furthermore,

$$card\, Imp[t - \tau(t), t] = card\, Imp[t + \rho - \tau(t + \rho), t + \rho] = m(t),$$

$t_i(t) \in Imp[t-\tau(t), t] \Rightarrow t_i(t+\rho) \in Imp[t+\rho-\tau(t), t+\rho]$ then $t_i(t), t_i(t+\rho) \in Imp[t-\tau(t), t+\rho]$,

$V(t_i(t + k\rho), x(t_i^-(t + k\rho))) \leq V(t_i(t), x(t_i^-(t))) + \nu\varepsilon$ for some $\nu \in \boldsymbol{R}^n$; $\forall k \in \boldsymbol{Z}_{0+}$

Then, for all $t \in \boldsymbol{R}_{0+}$, one has that the solution is almost ε-oscillatory with period ρ in the sense [4-5] that:

$$\left\| x(t + \rho) - x(t) \right\| \leq K \int_{-\tau(t)}^{0} \left\| x(s + t + \rho) - x(s + t) \right\| ds$$

$$\leq \varepsilon \left[K \left(\tau(t) + m(t) \max_{t_i(t) \in [t-\tau(t), t]} \left\| V\left(t_i(t), x\left(t_i^-(t)\right)\right) \right\| \right) + \|v\| \right] \tag{16}$$

provided that

$$\|v\| \leq 1 - K \sup_{t \in \mathbf{R}_{0+}} \left(\tau(t) + m(t) \max_{t_i(t) \in [t-\tau(t), t]} \left\| V\left(t_i(t), x\left(t_i^-(t)\right)\right) \right\| \right) \tag{17}$$

under the necessary condition that

$$K \leq \sup_{t \in \mathbf{R}_{0+}} \left(\tau(t) + m(t) \max_{t_i(t) \in [t-\tau(t), t]} \left\| V\left(t_i(t), x\left(t_i^-(t)\right)\right) \right\| \right)^{-1} \tag{18}$$

Note, on the other hand that $\tilde{x}(t^-) = x(t) - x(t^-) = K(t)x(t^-)$, where $K(t) = 0$ implies $\tilde{x}(t^-) = 0$ so that $x(t) \neq x(t^-)$ provided that $x(t^-) \neq 0$ if $K(t) \neq 0$. Then, for some $\sigma : \mathbf{R}_{0+} \to \mathbf{R}_{0+}$, one has

$$\left\| x(t^-) \right\| = \left\| \int_{-\tau(t)}^{0^-} F(s+t, x(s+t), V(s+t, x(s+t)))ds \right\| \leq \tau(t)\sigma(t^-) \sup_{s \in [t-\tau(t), t)} \| F(s, x(s), 0) \| \tag{19}$$

Note the following features:

a) if $\tilde{x}_t(s, x(s^-)) = 0$, i.e. $\tilde{x}(s, x(s^-)) = 0$ for $s \in [t - \tau(t), 0)$, equivalently, $Imp[t - \tau(t), t) = \varnothing$ then $\sigma(t^-) = 1$,

b) If $\sigma(t^-) < 1$ then there is a net vaccination impulsive control action on $[t - \tau(t), 0)$ with a net reduction of the infective population at the impulsive instants. In this case,

$$\left\| x(t^-) \right\| \leq \tau(t)\sigma(t^-) \sup_{s \in [t-\tau(t), t)} \| F(s, x(s), 0) \| \leq \tau(t) \sup_{s \in [t-\tau(t), t)} \| F(s, \sigma(t)x(s), 0) \| \tag{20}$$

$$\| x(t) \| \leq \tau(t)\sigma(t) \sup_{s \in [t-\tau(t), t]} \| F(s, x(s), 0) \| \leq \tau(t) \sup_{s \in [t-\tau(t), t]} \| F(s, \sigma(t)x(s), 0) \| \tag{21}$$

c) If $\sigma(t^-) > 1$ then there is a net increase of the infection at impulsive time instants on $[t - \tau(t), 0)$ with a net reduction of the infective population at the impulsive instants.

In a similar way, we obtain:

$$\| x(t) \| = \left\| \int_{-\tau(t)}^{0} F(s+t, x(s+t), V(s+t, x(s+t)))ds \right\| \leq \tau(t)\sigma(t) \sup_{s \in [t-\tau(t), t]} \| F(s, x(s), 0) \| \tag{22}$$

Assume that $\sigma(t^-) \leq 1$ and that $F(t, x(t^-), 0)$ is globally Lipschitz so that for some constant $L \in \mathbf{R}^+$, one gets under comparison with the impulsive-free zero solution:

$$\left\| x(t^-) \right\| \leq \tau(t) \sup_{s \in [t-\tau(t), t)} \left\| \sigma(t^-) F(s, x(s), 0) - F(s, 0, 0) \right\| \leq \tau(t)\sigma(t^-) L \sup_{s \in [t-\tau(t), t)} \| x(s) \|, \tag{23}$$

$$\| x(t) \| \leq \tau(t)\sigma(t) L \sup_{s \in [t-\tau(t), t]} \| x(s) \| \leq \tau(t)\sigma(t^-) L \sup_{s \in [t-\tau(t), t)} \left\| x(s^-) \right\| \tag{24}$$

If $\sigma(t^-) \le \frac{M}{\tau(t)\,L \sup\limits_{s\in[t-\tau(t),\,t)} \|x(s)\|}$ and $t \notin Imp[t-\tau(t),\,t]$ then $\|x(t)\| = \|x(t^-)\| \le M$ If instead $\sigma(t^-) > \frac{M}{\tau(t)\,L \sup\limits_{s\in[t-\tau(t),\,t)} \|x(s)\|}$ and $t = t_{m(t)} \in Imp[t-\tau(t),\,t]$ with $K(t)$ being such that $\sigma(t) \le \frac{\lambda M}{\tau(t)\,L \sup\limits_{s\in[t-\tau(t),\,t]} \|x(s)\|}$ for some $\lambda \in (0\,,1-\lambda_0)$ then $\|x(t)\| \le \lambda M < (1-\lambda_0)M$ and since $x(t)$ is continuous in $[t\,,t+\gamma_0)$ since it is time-differentiable in $(t\,,t+\gamma_0)$ for some, since it can be concluded that there exists some $\gamma \in [0\,,\gamma_0]$ such that $\|x(t)\| \le M$ on $\gamma_0 = \gamma_0(t) \in \boldsymbol{R}_+$ $[t\,,t+\gamma_0)$ provided that $\|x(t)\| \le M$ since $[t\,,t+\gamma) \cap Imp[t+\gamma-\tau(t+\gamma),\,t+\gamma) = \varnothing$. Now, proceed in the same way with $\{t+\gamma\} = Imp[t+\gamma-\tau(t+\gamma),\,t+\gamma]$ if $\sigma(t^- + \gamma) > \frac{M}{\tau(t+\gamma)\,L \sup\limits_{s\in[t-\tau(t),\,t)} \|x(s+\gamma)\|}$ so that $\sigma(t+\gamma) \le \frac{M}{\tau(t+\gamma)\,L \sup\limits_{s\in[t-\tau(t),\,t]} \|x(s+\gamma)\|}$ and $\|x(t+\lambda)\| \le \lambda M$ and $(t+\gamma) \notin Imp[t+\gamma-\tau(t+\gamma),\,t+\gamma]$ with $Imp[t+\gamma-\tau(t+\gamma),\,t+\gamma] = \varnothing$, otherwise. We can proceed in the same way along the whole time interval $[0\,,\infty)$.

4. Concluding remarks

In this paper, the properties of a new proposed integral model which might be related in particular cases to epidemic models has been discussed. This model incorporates delays and two coupled sub-states. The positivity, periodicity and almost-periodicity of the solution have being discussed under impulsive controls.

References

[1] B. Xu and R. Yuan, *Journal of Mathematical Analysis and Applications* **304**, 249-268 (2005).
[2] B. Xu and Q. Zhang, *Nonlinear Analysis* **74**, 5600-5605(2011).
[3] R. Torrejon, *Journal of Mathematical Analysis and Applications* **156**, 510-534(1991).
[4] S. Chen and R. Torrejon, *Nonlinear Analysis, Theory, Methods & Applications*, **27**, 863-877(1996).
[5] B. Xu and R. Yuan, *Journal of Mathematical Analysis and Applications* **304**, 249-268(2005).
[6] B. Xu and Z. Zhang, *Nonlinear Analysis* **71**, 141-148 (2009).
[7] H.S. Ding and G.M. N´Guerekata, *Applied Mathematics Letters* **26**, 881-885 (2013).
[8] M. De la Sen, R.P. Agarwal, A. Ibeas and S. Alonso-Quesada, *Advances in Difference Equations* **2011**, Article ID 48608 (2011).
[9] M. De la Sen, R.P. Agarwal, A. Ibeas and S. Alonso-Quesada, *Advances in Difference Equations* **2010**, Article ID 281612 (2010).
[10] A. Bilbao-Guillerna, M. De la Sen, A. Ibeas and S. Alonso-Quesada, *Discrete Dynamics in Nature and Society* **2005**, pp. 31-67 (2005).
[11] M. De la Sen, *IEEE Transactions on Automatic Control* **44**, pp. 597-602 (1999).
[12] M. Delasen, *IEEE Transactions on Automatic Control* **31**, pp. 582-586 (1986).
[13] M. Delasen, *IEE Proceedings- D Control Theory and Applications* **131**, pp. 146-157 (1984).

Computational Modelling and Optimal Control of Ebola Virus Disease with non-Linear Incidence Rate

[1]Takaidza I., [2]Makinde O. D., [3]Okosun O. K.

[1]School of Information Technology, North West University, Vanderbijlpark, RSA
[2]Faculty of Military Sciences, Stellenbosch University, Saldanha, RSA
[3]Mathematics Department, Vaal University of Technology, Vanderbijlpark, RSA

Isaac.Takaidza@nwu.ac.za

Abstract. The 2014 Ebola outbreak in West Africa has exposed the need to connect modellers and those with relevant data as pivotal to better understanding of how the disease spreads and quantifying the effects of possible interventions. In this paper, we model and analyse the Ebola virus disease with non-linear incidence rate. The epidemic model created is used to describe how the Ebola virus could potentially evolve in a population. We perform an uncertainty analysis of the basic reproductive number R_0 to quantify its sensitivity to other disease-related parameters. We also analyse the sensitivity of the final epidemic size to the time control interventions (education, vaccination, quarantine and safe handling) and provide the cost effective combination of the interventions.

1. Introduction

The 2014 Ebola virus outbreak is the largest and most complex ever to occur since the disease's first appearance in 1976 via simultaneous outbreaks in Nzara, Sudan and in Yambuku, Democratic Republic of Congo. The latter occurred in a village next to Ebola River, from where the disease took its name. The Ebola virus is a zoonotic filovirus which erupts occasionally and has caused at least 14 confirmed outbreaks in Africa between 1976 and 2006 [13]. The exact reservoir of Ebola viruses is still unknown, but researchers believe the most likely natural hosts are fruit bats [7]. Ebola causes haemorrhagic fever and death in humans after about ten days, and people in contact with infectives can be infected. The virus can be spread from one person to others through direct contact with blood or body fluids (urine, saliva, sweat, faeces, vomit, breast milk, and semen) of a person who is sick with or has died from Ebola and objects (like needles and syringes) that have been contaminated with body fluids [3].

The typical trajectory of the disease begins with an average incubation period of one to two weeks. Patients mostly present with fever, asthenia, diarrhoea, abdominal pain, headache, arthralgia, myalgia, sore throat, dysphagia, and conjunctivitis [19]. One week after the onset of symptoms a rash often appears followed by haemorrhagic complications, leading to death after an average of ten days in 50–90% of infections. The virus can lurk in the body for more than a week before it begins a cascading meltdown of the immune system, blood vessels and vital organs [18]. Recovery usually occurs in two weeks to two months after the onset of symptoms. Most individuals acquire infection after direct contact with blood, bodily secretions and tissues of infected ill or dead humans and non-human primates [8, 15]. There is evidence that individuals (health-care workers, relatives) may become infected following contacts with patients' body fluids or direct contact with patients during a visit at the hospital or participation in traditional burial ceremonies [11]. Ebola is not transmissible if someone is asymptomatic or once someone has recovered from it. However, the virus has been found in semen for up to three months [7]. Vaccine and therapeutic strategies have been under development since around 2003 [17, 6].

There is as yet no proven treatment for Ebola virus disease (EVD). However, a range of potential treatments including blood products, immune therapies and drug therapies are currently being investigated. No licensed vaccines are available yet, but two potential vaccines are undergoing human safety tests (WHO). The recent outbreak has had more cases and deaths than all previous ones combined. More than 20 000 people had been infected with Ebola, as of 28 December 2014, as the recent outbreak continued to spread globally. Specific statistics were as follows: Liberia (8018 cases, 3423 deaths), Sierra Leone (9446 cases, 2758 deaths), Guinea (2707 cases, 1709 deaths), Nigeria (20 cases, 8 deaths), Mali (8 cases, 6 deaths), United States (4 cases, 1 death), Senegal (1 case, 0 deaths) and Spain (1 case, 0 deaths) [3].

The authors in [2] showed, using SIR and SEIR models, that it was possible to simulate Ebola outbreaks. They considered the 1976 outbreak in Yambuku, DRC (then Zaire) and the 1995 outbreak in Kikwit, DRC. Chowell et al 2004 modelled the course of the Congo 1995 and Uganda 2000 Ebola outbreaks via an SEIR (susceptible-exposed-infectious-removed) epidemic model that included a smooth transition in the transmission rate after control interventions are put in place. The control measures implemented during these two outbreaks (including education and contact tracing followed by quarantine) reduced the final epidemic size by a factor of 2 relative to the final size with a 2-week delay in their implementation.

Previously published Ebola Haemorrhagic Fever (EHF) data from two epidemics was analysed using a stochastic compartmental model which incorporated the explicit settings of the transmission in the community, in the hospital and during burial ceremonies [13]. The goal was to better understand and to provide insight into where control interventions could be targeted in the future. The infectious phase was subdivided into three stages to account for transmission in the community, in the hospital (including isolation wards), and after death during traditional burial. Various epidemic scenarios were simulated to explore the impact of control interventions on a potential epidemic. Increasing hospitalization rate, for both epidemics, reduced the predicted epidemic size.

Yarus [21] analysed an epidemic model consisting of susceptible-infected-recovered-dead created from first order differential equations to describe how the Ebola virus could potentially ravage a population. Rivers et al 2014 used existing data from Liberia and Sierra Leone to parameterize a mathematical model of Ebola and used this model to forecast the progression of the epidemic, as well as the efficacy of several interventions, including increased contact tracing, improved infection control practices, the use of a hypothetical pharmaceutical intervention to improve survival in hospitalized patients. The Ebola virus epidemic was described using an SEIR model to estimate the reproduction number in the West Africa outbreak [1]. The maximum likelihood estimates of the basic reproduction number are 1.51 for Guinea, 2.53 for Sierra Leone and 1.59 for Liberia.

A discrete, stochastic SEIR model was combined with a two-scale community network model to demonstrate that the different regional trends may be explained by different community mixing rates [12]. The spatio-temporal epidemiological modeler (STEM) developers implemented a preliminary epidemiological model, for the three African countries mostly affected by the Ebola epidemic, based on 8 compartments and 8 differential equations. The model includes the usual susceptible, exposed, infectious states along with 5 additional compartments: death (but not buried), buried, clinical isolation, and burial. The usual recovered or removed compartment is represented by asymptomatic but still infectious individuals [9].

The paper is organized as follows. In Section 2, we present a compartmental deterministic model consisting of ordinary differential equations describing the transmission dynamics of the Ebola virus disease given the underlying assumptions and also provide the basic properties of the model as well as investigate the existence and stability of equilibria. Section 3 is devoted to optimal control of the disease making use of Pontryagin's maximum principle. In Section 4, we present and discuss the numerical simulations results. Cost effectiveness analysis is the subject of Section 5 and our conclusions are in Section 6.

2. Model Formulation

The population, $N(t)$, is sub-divided into the following non-overlapping but exhaustive compartments: the susceptible, $S(t)$; the exposed, $E(t)$; the infected, $I(t)$; individuals on treatment, $T(t)$, the recovered, $R(t)$ and those who die due to the disease $D(t)$. We assume that individuals are recruited into the population being susceptible, through birth and immigration, at a constant rate Q. The disease is transmitted to the susceptible by the infected, those on treatment and the deceased corpses at rates β_1, β_2 and β_3 respectively, $\beta_1 > \beta_3 > \beta_2$ due to the effect of treatment and that the virus in a corpse is not as deadly. Individuals in the exposed class progress to the infected class at a rate σ, the infected migrate to the treatment class at a rate γ and the treated recover at rate b. Individuals in all compartments die due to natural causes at a rate μ while the infected and the treated may also die due to the Ebola virus at rates m_1 and m_2 respectively, $m_1 > m_2$. The recovered individuals progress to the susceptible class at a rate α while the deceased are removed from the system through safe burial at a rate ρ. We denote by u_1 the control measure due to quarantine of the exposed and infected individuals as well as safe handling of the dead while u_2 represents the control parameter due to efficacy of the treatment drug used for Ebola virus victims.

The resulting ***state system*** of first order differential equations is the following:

$$\frac{dS}{dt} = Q - (1 - u_1)\lambda S + \alpha R - \mu S$$

$$\frac{dE}{dt} = (1 - u_1)\lambda S - (\sigma + \mu)E$$

$$\frac{dI}{dt} = \sigma E - (\gamma + m_1 + \mu)I$$

$$\frac{dT}{dt} = \gamma I - (1 + u_2)bT - m_2 T - \mu T$$

$$\frac{dR}{dt} = (1 + u_2)bT - (\alpha + \mu)R$$

$$\frac{dD}{dt} = m_1 I + m_2 T - \rho D$$

where $\lambda = \frac{\beta_1 I + \beta_2 T + \beta_3 D}{a + I + T + D}$ is the force of infection.

2.1. Basic Model Properties

The model represents the dynamics of a human population. Hence, all its parameters and initial conditions are assumed to be non-negative. Similarly, for the model to make sense, all its solutions must be non-negative. Thus, we claim the following:

Theorem: The solutions $S(t)$, $E(t)$, $I(t)$, $T(t)$, $R(t)$ and $D(t)$ of (1) with non-negative initial conditions exist for all $t > 0$, and are unique. Furthermore, the solutions are non-negative and are bounded for all $t > 0$.

Proof: From the first equation of the ***state system***,

$$\frac{dS}{dt} \geq -(\lambda + \mu)S$$

Assuming $S(0) \geq 0$ and that all other parameters are non-negative, we have

$S(t) \geq S(0)exp\left(-\int_0^t (\lambda + \mu)dv\right) \geq 0$ for all $t > 0$.

Similarly, $E(t)$, $I(t)$, $T(t)$, $R(t)$ and $D(t)$ are non-negative for $t > 0$.

Now, the total population, $N(t)$ satisfies

$$Q - (\mu + \rho)N \leq \frac{dN}{dt} = Q - \mu N - \rho D \leq Q - \mu N$$

This implies that

$$\frac{Q}{\mu + \rho} \leq N(t) \leq \frac{Q}{\mu}$$

□

The set

$$\Gamma = \left\{(S, E, I, T, R, D) \in \mathbb{R}_+^6 : S + E + I + T + R + D = N \leq \frac{Q}{\mu}\right\}$$

is positively invariant and attracting with respect to the model. The model is thus well-posed epidemiologically and mathematically, hence we can study the qualitative properties of the system.

2.2. Analysis of the Model

The model has an Ebola-free equilibrium (EFE) given by

$$\xi^0 = (S^0, E^0, I^0, T^0, R^0, D^0) = \left(\frac{Q}{\mu}, 0, 0, 0, 0, 0\right)$$

The transmissibility of an infection can be quantified by its basic reproductive number R_0, defined as the mean number of secondary infections seeded by a typical infective into a completely susceptible (naive) host population [5]. Thus, to determine the impact of introducing an infectious individual in a population of completely susceptible individuals, we proceed by computing the basic reproductive number of the system. This is the spectral radius of the next generation matrix $K = F V^{-1}$, where F and V respectively represent the transmission (at the EFE) and transition matrices and are given by

$$F = \begin{bmatrix} 0 & f_1 & f_2 & f_3 \\ 0 & 0 & 0 & 0 \\ 0 & 0 & 0 & 0 \\ 0 & 0 & 0 & 0 \end{bmatrix}; V = \begin{bmatrix} \sigma+\mu & 0 & 0 & 0 \\ -\sigma & \gamma+m_1+\mu & 0 & 0 \\ 0 & -\gamma & v & 0 \\ 0 & -m_1 & -m_2 & \rho \end{bmatrix}$$

where $f_i = \frac{(1-u_1)\beta_i Q}{a\mu}$, $i = 1,2,3$ and $v = (1+u_2)b + \mu + m_2$. F describes the production of new infections whereas V describes the changes in states. It follows that, at the EFE, the basic reproduction number is

$$R_0 = \frac{Q(1-u_1)[\rho\beta_1((1+u_2)b+\mu+m_2)+\gamma(\rho\beta_2-m_2\beta_3)]}{a\mu\rho(\sigma+\mu)(\gamma+m_1+\mu)((1+u_2)b+\mu+m_2)}$$

Lemma: The Ebola free equilibrium, ξ^0, of the model represented by the ***state system*** is locally asymptotically stable (LAS) if $R_0 < 1$, and unstable if $R_0 > 1$.

The Jacobian matrix J of the **state system** at ξ^0 is

$$J(\xi^0) = \begin{bmatrix} -\mu & 0 & -f_1 & -f_2 & \alpha & -f_3 \\ 0 & -y & f_1 & f_2 & 0 & f_3 \\ 0 & \sigma & -u & 0 & 0 & 0 \\ 0 & 0 & \gamma & -v & 0 & 0 \\ 0 & 0 & 0 & w & -x & 0 \\ 0 & 0 & m_1 & m_2 & 0 & -\rho \end{bmatrix}$$

where $y = \sigma + \mu$, $u = \delta + m + \mu$, $w = (1-u_2)b$ and $x = \alpha + \mu$. The eigenvalues are

$$\Lambda_1 = 0 = \Lambda_2,\ \ \Lambda_3 = -\mu < 0, \quad \Lambda_4 = -\rho < 0,$$

$$\Lambda_5 = -\frac{1}{2}\left(\sigma + 2\mu + \gamma + m_1 + \sqrt{\frac{4\sigma(1-u_1)\beta_1 Q + a\mu(\sigma+2\mu+\gamma+m_1)^2}{a\mu}}\right) < 0$$

$$\Lambda_6 = -\frac{1}{2}\left(\sigma + 2\mu + \gamma + m_1 - \sqrt{\frac{4\sigma(1-u_1)\beta_1 Q + a\mu(\sigma+2\mu+\gamma+m_1)^2}{a\mu}}\right)$$

$\Lambda_6 < 0$ only if $\sigma + 2\mu + \gamma + m_1 - \sqrt{\frac{4\sigma(1-u_1)\beta_1 Q + a\mu(\sigma+2\mu+\gamma+m_1)^2}{a\mu}} > 0$

implying

$$0 > 4\sigma(1-u_1)\beta_1 Q$$

only if $R_0 < 1$. Thus, ξ^0 is LAS.

The state system has an endemic equilibrium $\xi^* = (S^*, E^*, I^*, T^*, R^*)$ described by

$$S^* = \frac{(\delta+\mu)E^*}{(1-u_1)\lambda^*};\ I^* = \frac{\delta E^*}{\delta+m+\mu};\ T^* = \frac{\gamma\delta E^*}{(\delta+m+\mu)[(1-u_2)b+\mu]};\ R^* = \frac{(1-u_2)b\gamma\delta E^*}{(\alpha+\mu)(\delta+m+\mu)[(1-u_2)b+\mu]}$$

where λ^* represents the endemic force of infection.

2.3. *Sensitivity Analysis*

The following partial derivatives of the reproduction number with respect to its parameters

$$\frac{\partial R_0}{\partial \beta_1}=\frac{(1-u_1)Q}{a\mu(\delta+\mu)};\ \frac{\partial R_0}{\partial Q}=\frac{(1-u_1)\beta_1}{a\mu(\delta+\mu)};\ \frac{\partial R_0}{\partial \delta}=-\frac{\beta_1(1-u_1)Q}{a\mu(\delta+\mu)^2};\ \frac{\partial R_0}{\partial \mu}=-\frac{\beta_1(1-u_1)Q(\delta+2\mu)}{a\mu^2(\delta+\mu)^2}$$

show that increasing (decreasing) β_1 or Q may increase (decrease) R_0 whilst increasing (decreasing) δ or μ may decrease (increase) R_0. To determine the relative importance of each parameter we make use of the normalized forward sensitivity index defined as follows:

Definition: The normalized forward sensitivity index of a variable U, that depends differentiably on a parameter p, is defined as:

$$\zeta_p^U=\frac{\partial U}{\partial p}\times\frac{p}{U}$$

We evaluate the sensitivity indices using the baseline parameter values given in Table 2.

Table 1: Sensitivity indices

Parameter	Index
β_1	1
Q	1
δ	-5.188755252
μ	-1.44800465

It turns out that the reproduction number, R_0, is most sensitive to the rate of progression of the exposed to the infected class, δ. In particular, a 1 % decrease (increase) in δ results in a 5.2 % increase (decrease) in R_0.

3. Optimal Control

To investigate the optimal level of efforts that would be needed to control the disease, we wish to minimize the number of exposed, infected and dead individuals, that is the classes E, I and D, and the cost of applying the controls u_1 and u_2 over a finite time interval $[0, T_f]$. We achieve this by defining an objective functional, J, by choosing a quadratic cost on the controls

$$J=\int_0^{T_f}(kE+nI+pD+b_1u_1^2+b_2u_2^2)dt$$

where k, n, p, b_1 and b_2 are positive weights. The weights of state variables are assigned depending on their relative importance while those of controls are assigned relative to their cost implications.

With the given objective functional, $J(u_1, u_2)$, our goal is to minimize the number of the exposed E and the infectious I while minimizing the cost of the controls $u_1(\mathrm{t})$ and $u_2(\mathrm{t})$. We thus seek an optimal control pair (u_1^*, u_2^*) such that

$$J(u_1^*, u_2^*) = \min\{J(u_1, u_2): (u_1, u_2) \in \mathcal{U}\}$$

where

$$\mathcal{U} = \{(u_1, u_2): u_i \text{ measurable with } 0 \le u_i \le w_i < 1, \mathrm{i} = 1, 2\,;\ \mathrm{t} \in [0, T_f]\}$$

is the control set where w_i represents the maximum coverage for control u_i.

The necessary conditions that an optimal pair $(\boldsymbol{u}^*; \boldsymbol{x}^*)$, where $\boldsymbol{u}^* = (u_1^*, u_2^*)$ and $\boldsymbol{x}^* = (S^*, E^*, I^*, T^*, R^*, D^*)$, must satisfy come from Pontryagin's Maximum Principle [16]. We use this principle to convert the problem of minimization of the objective functional coupled with the state variables into a problem of minimizing point-wise a Hamiltonian, H, with respect to the controls u_1 and u_2.

$$H = kE + nI + pD + b_1 u_1^2 + b_2 u_2^2 + \lambda_S \cdot \frac{dS}{dt} + \lambda_E \cdot \frac{dE}{dt} + \lambda_I \cdot \frac{dI}{dt} + \lambda_T \cdot \frac{dT}{dt} + \lambda_R \cdot \frac{dR}{dt} + \lambda_D \cdot \frac{dD}{dt}$$

where λ_S, λ_E, λ_I, λ_T, λ_R and λ_D are adjoint or co-state variables. By applying Pontryagin's Maximum Principle and the existence result for the optimal control [10], we obtain the following:

Proposition: *For the optimal control pair (u_1^*, u_2^*) that minimizes $J(u_1, u_2)$ over $\mathcal{U}$, there exist adjoint variables λ_S, λ_E, λ_I, λ_T, λ_R and λ_D satisfying the following:*

(i) Adjoint System

$$\frac{d\lambda_S}{dt} = (1 - u_1)\left(\frac{\beta_1 I + \beta_2 T + \beta_3 D}{a + I + T + D}\right)(\lambda_S - \lambda_E) + \mu\lambda_S$$
$$\frac{d\lambda_E}{dt} = -k + \sigma(\lambda_E - \lambda_I) + \mu\lambda_E$$
$$\frac{d\lambda_I}{dt} = -n + B_1(\lambda_S - \lambda_E) + \gamma(\lambda_I - \lambda_T) + m_1(\lambda_I - \lambda_D) + \mu\lambda_I$$
$$\frac{d\lambda_T}{dt} = B_2(\lambda_S - \lambda_E) + (1 + u_2)b(\lambda_T - \lambda_R) + m_2(\lambda_T - \lambda_D) + \mu\lambda_T$$
$$\frac{d\lambda_R}{dt} = \alpha(\lambda_R - \lambda_S) + \mu\lambda_R$$
$$\frac{d\lambda_D}{dt} = -p + B_3(\lambda_S - \lambda_E) + \rho\lambda_D$$

where $B_i = \frac{(1-u_1)S\{(a+I+T+D)\beta_i - (\beta_1 I + \beta_2 T + \beta_3 D)\}}{(a+I+T+D)^2}$, $i = 1, 2, 3$.

(ii) Transversality Conditions

$$\lambda_S(T_f) = \lambda_E(T_f) = \lambda_I(T_f) = \lambda_T(T_f) = \lambda_R(T_f) = \lambda_D(T_f) = 0$$

(iii) Stationary Values

$$u_1^* = min\left\{w_1, max\left\{0, \frac{(\beta_1 I + \beta_2 T + \beta_3 D)S(\lambda_E - \lambda_S)}{2(a + I + T + D)b_1}\right\}\right\}$$

$$u_2^* = min\left\{w_2, max\left\{0, \frac{bT(\lambda_R - \lambda_T)}{2b_2}\right\}\right\}$$

4. Numerical Simulations

We study numerically an optimal transmission parameter control for the Ebola virus model. The optimal control set is obtained by solving the associated optimality system which consists of ***state*** and ***adjoint*** systems. An iterative fourth order Runge-Kutta scheme is used for solving the optimality system. We start to solve the state equations with a guess for the controls over the simulated time. The state equations are solved using a forward method with given initial conditions for the state variables. The corresponding adjoint system is solved using a backward scheme with the ***transversality conditions***. The controls are updated by using a convex combination of the previous controls and the ***stationary values***. This process is repeated and iterations stopped if the values of the unknowns at the previous iterations are very close to the ones at the present iterations [14].

Table 2: Model parameters

Parameter	Description	Value
Q	Recruitment rate	10
β_2	Contact rate between the susceptible and the treated	0.112195104
β_1	Contact rate between the susceptible and the infected	0.224390208
β_3	Contact rate between the susceptible and the dead	0.168292656
γ	Migration rate for infected to treatment	0.12195120
α	Progression rate for recovered to susceptible	0.023
b	Recovery rate for the treated	0.03560976
m_1	Ebola-induced mortality rate for the infected	0.708
m_2	Ebola-induced mortality rate for the treated	0.508
μ	Natural death rate	0.0863414496
σ	Progression rate for exposed to infected	0.1063830
ρ	Removal rate for the dead	0.5

For the simulations, we choose the model parameter values as in Table 1. All the values except the recruitment rate Q, which is assumed, are sourced from [9]. We assume that the weight factor b_1, associated with control u_1, is lower than b_2, which is associated with the control u_2. This assumption is based on the fact that it probably costs more to treat an infected person. Moreover, the cost associated with treatment will also include the cost of medical examinations, any prescribed drugs and hospitalization, thus making it costlier. We make use of the following objective functional parameter values: $n = 75$, $k = 150$, $p = 50$, $b_1 = 50$, $b_2 = 120$. For illustrative purposes we make use of $w_1 = 0.9 = w_2$ and the initial state conditions $S(0) = 100$, $E(0) = 30$, $I(0) = 2$, $T(0) = 1$, $R(0) = 0$ and $D(0) = 0$.

We discuss results for the case when deploying both the quarantine and treatment interventions. Numerical results show that there is a marked reduction from the onset in the number of the exposed when employing interventions, Fig. 1. Interventions result in a gradual decline in the infected class, Fig. 2. The declines in the treated and recovered classes, Fig. 3 and Fig. 4, are not marked. This may be due to the fact that fewer people are being exposed to the infection.

IOP Conf. Series: Journal of Physics: Conf. Series **818** (2017) 012003 doi:10.1088/1742-6596/818/1/012003

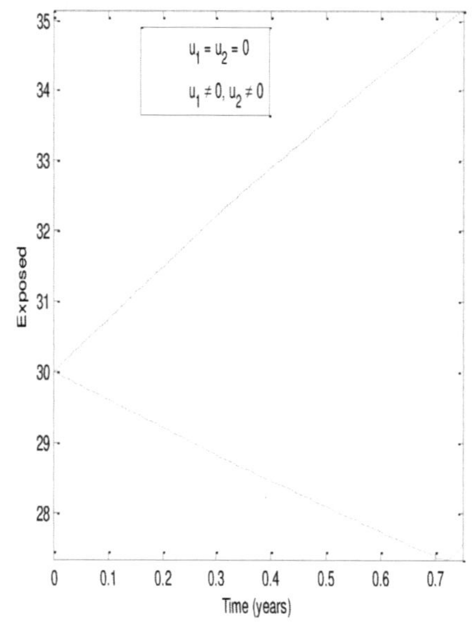

Fig. 1: Exposed

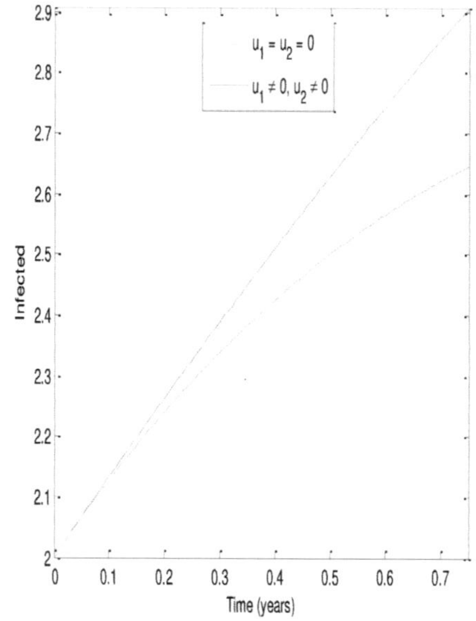

Fig. 2: Infected

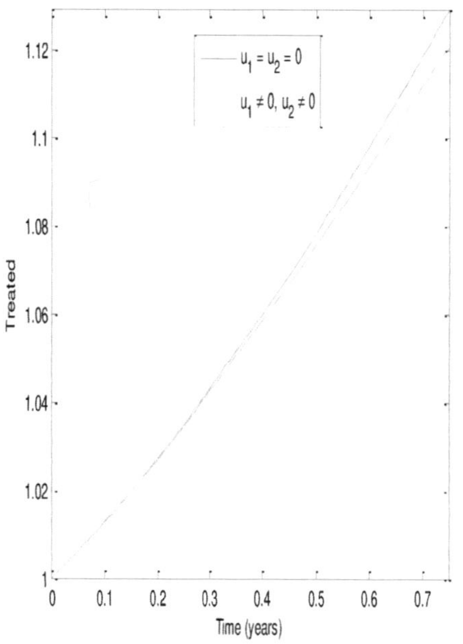

Fig. 3: Treated

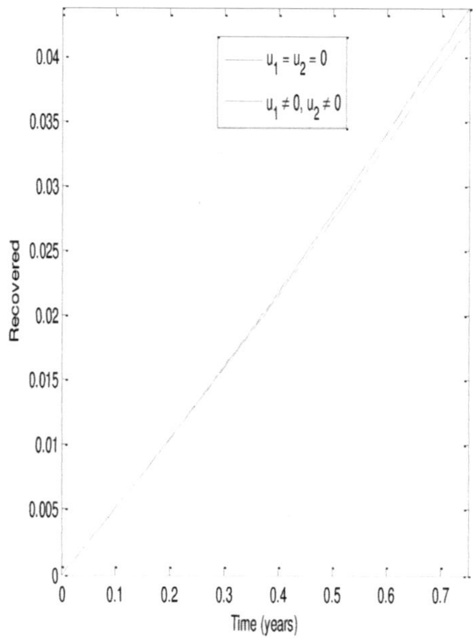

Fig. 4: Recovered

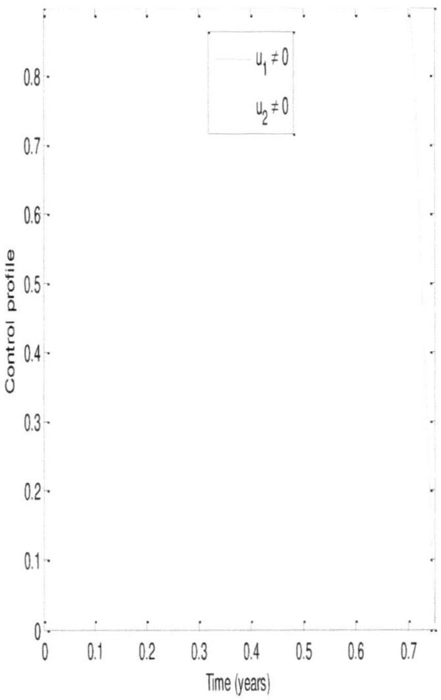

Fig. 5: Control Profiles

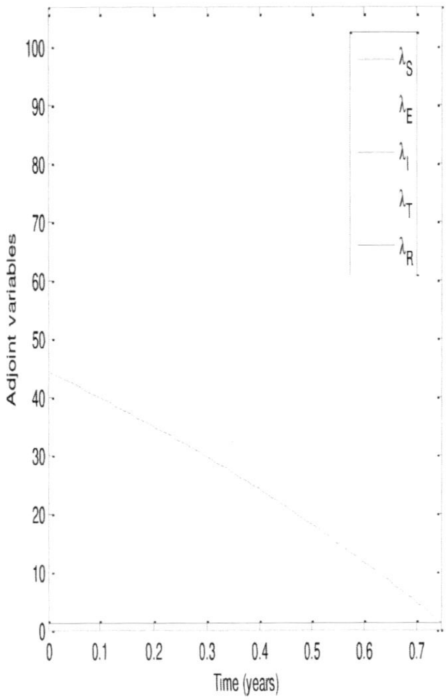

Fig. 6: Adjoint Variables

Control profiles, Fig. 5, show that maximum effort needs to be put in deploying quarantine measures (for about 8 months if the intervention period is 9 months) to ensure that the exposed and infected do not interact with the rest of the population. The treatment control can then be minimal. Shadow prices, Fig. 6, show that the exposed and the infected are the costliest.

It is of utmost importance to determine the cost effective interventions so that funds may be appropriately spent given the tight budgets national economies operate on.

5. Cost Effectiveness

A critical goal of public health programmes is the realization of positive impact for a given population. This impact is usually a result of a combination of various inputs and it is often not due to a single intervention. Cost effective analysis is one economic evaluation tool used to measure the costs and consequences of alternative programmes. The measures are then compared to determine how the greatest health benefits can be generated. We use indicators related to a change in health status such as the estimated number of fatalities or infections averted to measure impact. The strategy which realises the most positive impact is identified by use of incremental cost effectiveness ratios (ICERs) defined by

$$ICER = \frac{Difference\ in\ costs\ between\ strategies}{Difference\ in\ health\ effects\ between\ strategies}$$

We consider as health effects the cases averted in the E and I classes for the following three strategies: A – quarantine measures; B – treatment measures; C – both controls employed. The strategies are ranked from the least effective by considering health effects and then compared pairwise using ICERs.

Table 3. Cases averted and associated costs

Strategy	Cases Averted	Costs
B	137	11 831
A	403	753.80
C	404	753.84

The ICER between B and A is given by $\frac{753.80-11\,831}{403-137} = -41.64$. It costs 41.64 less for each additional case averted from B to A, so B is excluded.

Next, we calculate the ICER between A and C, which is 0.04. Hence, it costs 0.04 more for each additional case averted as we switch from A to C. So, we exclude C and come to that strategy A is the most cost effective.

6. Conclusion

Shadow prices show that the cost and impact of the exposed is very high, followed by the infected. This may result in negative effects on the population. The results suggest that quarantining of the exposed and the infected is the most effective strategy. However, the marginal additional cost per extra case averted when switching from strategy A to C justifies the inclusion of treatment measures. Therefore, budgetary provision needs to be made to include the treatment of the Ebola virus victims. Control programs that follow these strategies can effectively reduce the incidence and endemicity of the Ebola virus disease.

References

[1] Althaus CL 2014 Estimating the reproduction number of Ebola virus during the 2014 outbreak in West Africa *PLOS Currents Outbreaks* doi:10.1371/currents.outbreaks.91afb5e0f279e7f29e7056095255b288.
[2] Astacio J, Briere D, Guillen M, Martinez, J, Rodriguez F and Valenzuela-Campos N 1996 Mathematical Models to study the Outbreaks of Ebola, *Biometrics Unit Technical Report*
[3] Centres for Disease Control and Prevention 2014 Review of Human-to-Human Transmission of Ebola Virus retrieved from http://www.cdc.gov/vhf/ebola/transmission/human-transmission.html
[4] Chowell G, Hengartner, NW, Castillo-Chavez C, Fenimore PW and Hyman JM 2004 The basic reproductive number of Ebola and the effects of public health measures: the cases of Congo and Uganda *Journal of Theoretical Biology* **229**(1):119 – 126
[5] Cintron-Arias A, Castillo-Chavez C, Bettencourt LMA, Lloyd AL and Banks HT 2009 Estimation of the effective reproductive number from disease outbreak data. *Mathematical Biosciences and Engineering* **Vol 6 No 2** 261 – 282 doi.10.3934/mbe.2009.6.261
[6] Clarke T and Knight J 2003 Fast vaccine offers hope in battle with Ebola *Nature* **424**: 602.
[7] CNN Library. 2014. Ebola fast facts
[8] Dowell SF, Mukunu R, Ksiazek TG, Khan AS, Rollin PE, Peters CJ 1999 Transmission of Ebola haemorrhagic fever: a study of risk factors in family members, Kikwit, Democratic Republic of the Congo, 1995. Commission de Lutte contre les Epidemies a Kikwit *Journal of Infectious Diseases* **179** (Suppl. 1): S87–S91.
[9] Eclipse https://wiki.eclipse.org/Ebola_Models
[10] Fleming WH and Rishel RW 1975 *Deterministic and Stochastic Optimal Control* (Springer

Verlag, New York)

[11] Hewlett BS. and Amola RP 2003 Cultural contexts of Ebola in northern Uganda *Emerging Infectious Diseases* **9**: 1242 – 48.

[12] Kiskowski MA 2014 Description of the Early Growth Dynamics of 2014 West Africa Ebola Epidemic *arXiv [q-bio.PE]*

[13] Legrand J, Grais RF, Boelle PY, Valleron AJ and Flahault A 2007 Understanding the dynamics of Ebola epidemics. *Epidemiol. Infect.* **135** (4), 610–621

[14] Lenhart S and Workman JT 2007 *Optimal Control Applied to Biological Models* (Chapman and Hall)

[15] Peters CJ and LeDuc JW 1999 An introduction to Ebola: the virus and the disease *Journal of Infectious Diseases* **179** (Suppl. 1): ix–xvi

[16] Pontryagin LS, Boltyanskii VG, Gamkrelidze RV and Mishchenko EF 1962 *The Mathematical Theory of Optimal Processes* (Wiley, New York)

[17] Rouquet P, Froment JM, Bermejo M, Kilbourn A, Karesh W, Reed P, Kumulungui B, Yaba P, Delicat A, Rollin PE and Leroy EM 2005 Wild animal mortality monitoring and human Ebola outbreaks, Gabon and Republic of Congo, 2001–2003. *Emerging Infectious Diseases* **11**: 283–290.

[18] The Washington Post. 2014. Ebola's catastrophic effect on the body

[19] World Health Organization 1978 Ebola haemorrhagic fever in Sudan, 1976. Report of a WHO/International Study Team *Bulletin of the World Health Organization* **56**: 247–270.

[20] World Health Organization 1978 Ebola haemorrhagic fever in Zaire, 1976. Report of an International Convention *Bulletin of the World Health Organization* **56**: 271–293.

[21] Yarus Z 2012 A Mathematical Look at the Ebola Virus

User Requirements Analysis For Digital Library Application Using *Quality Function Deployment.*

Lily Wulandari[1)] , Lana Sularto[2)], Tristyanti Yusnitasari[3)], Diana Ikasari[4)]
[1,3,4] Faculty of Computer Science, Gunadarma University
[2] Faculty of Economics, Gunadarma University
Email : lily@staff.gunadarma.ac.id, lana@staff.gunadarma.ac.id, tyusnita@staff.gunadarma.ac.id, d_ikasari@staff.gunadarma.ac.id

Abstract

This study attemp to build Smart Digital Library to be used by the wider community wherever they are. The system is built in the form of Smart Digital Library portal which uses semantic similarity method (Semantic Similarity) to search journals, articles or books by title or author name. This method is also used to determine the recommended books to be read by visitors of Smart Digital Library based on testimony from a previous reader automatically. Steps being taken in the development of Smart Digital Library system is the analysis phase, design phase, testing and implementation phase. At this stage of the analysis using WebQual for the preparation of the instruments to be distributed to the respondents and the data obtained from the respondents will be processed using Quality Function Deployment. In the analysis phase has the purpose of identifying consumer needs and technical requirements. The analysis was performed to a digital library on the web digital library Gunadarma University, Bogor Institute of Agriculture, University of Indonesia, etc. The questionnaire was distributed to 200 respondents. The research methodology begins with the collection of user requirements and analyse it using QFD. Application design is funded by the government through a program of Featured Universities Research by the Directorate General of Higher Education (DIKTI). Conclusions from this research are identified which include the Consumer Requirements of digital library application. The elements of the consumers requirements consists of 13 elements and 25 elements of Engineering Characteristics digital library requirements. Therefore the design of digital library applications that will be built, is designed according to the findings by eliminating features that are not needed by restaurant based on QFD House of Quality.

Keywords: systems analysis, applications design, smart digital library, quality function deployment.

1. Introduction

The main activity of the digital library is a data search resources in the form of books, journals or articles that represented by the title of each of these resources. When the authors determine the choice of terminology which represents the title is very influenced by the background knowledge and ability. Differences in knowledge and ability results in choosing terminology for titles of books, journals and articles that represent the contents of these writings. The use of different terminology does not mean that the contents of books, journals or articles are very different. This is because the two terms are semantically similar (have the same meaning) although lexically different. Therefore, the search method classic library of digital resources (keyword matching search-based) will fail to find resources that are semantically similar but lexically different. This failure is characterized by the irrelevance of the search results. This is exacerbated by the fact that the submitted query does not fairly represent the resource that people wants to find. Usually done by doing query repeatedly by changing the form of selection query terminology. The search method of digital library resources that can be automatically enriched by exploiting the semantic heterogeneity is very important to identify the terminology and broaden the search results. Some search techniques have been developed, such as full text search, metadata search, and search with semantic networks. Based on observations made on several digital library sites, search by title has not been made using techniques that based on common semantics search, using common semantics technique search will results of the scope of data will be obtained wider, if using keyword based technique, to get the information it must be made multiple times query, because of the book's title can be expressed in terms that were different despite having a similarity of meaning. Examples of search with the keyword "Programming" actually has in common with the word "Coding" or "Making Program Code". So the search by considering the semantic can save queries made with a lot of or extensive data coverage.

In addition to standard digital library research facility, this research also trying to add a new function which is actually a function already exists in the online book store that is book recommendations. People can not see a good book based on the title of the book, sometimes the titles do not talk about the contents of the book. Testimony is included in the opinion mining or sentiment analysis, which is a process of understanding, extract and process the textual data automatically to get the sentiment of information contained in an opinion sentence. Turney research in 2002, namely the opinion mining by using data contains consumer reviews data on a product. The method used in this study is the semantic orientations using pointwise Mutual Information (SO-PMI)[17]. Sentiment analysis is a combination of text mining and natural processing. Several methods that can be used to solve the problem of opinion mining is Naive Bayes classifier (NBC). NBC used to classify the opinions into positive opinions and negative opinions. Another method that can be used is KNN Clustering, NN, KMean, and Support Vector Machine (SVM) Clustering. Usually the book will be a best seller after people read later testified against the book. Testimonials will give rise to the interest to read the book if the result is well worth the testimony or recommend that the book is a good read. In general recommendation made by testimonials for books or other products in the online or digital media library is only done by selecting from a few words that represent the testimony of each user or reader. Such conditions can not give users flexibility in expressing his opinion on a book or any other product as testimonials that do generally, which is a sentence not a word choice. The difficulty in measuring the ranking of a book based on testimony is the diversity of representation at the time of making the testimony by users even though the meaning of the testimony has the same meaning.

Ranking can also be done on the website of the hotel, resort, tourist attractions and others. Such websites usually provide features to provide recommendations on product reviews or just consumer reviews. As examples of websites that have the product review is www.amazon.com, www.download.com, www.yelp.com, Website yelp.com has a feature for users to give testimony to the product and then to a ranking done by giving a checklist on the asterisk, and its website amazon.com just to give you a checklist on the asterisk by it can not be known whether the product is the result of reading a book is good or bad. In this case it can be said that the lack of response or comment from the user, it could be when the user clicks or to cite a book, the book turned out to be not in accordance with what is expected or desired. Measurement rankings based on testimony can be done manually by a reviewer read testimonials. then classify it after that do the rankings, but this process requires substantial time, is expected to have a tool to automatically measure the rank of books or other products based on such testimony. Making a Smart Digital Library is expected to be very useful for readers to search information, where Smart Digital Library contains a title search based on common semantics (semantic similarity) and advice books are widely read by testimonials from readers of the book is done by looking at the similarities semantics.

Sample data for ranking the books based on testimonies by similarity semantic taken from a catalog of books on the website and also the physical book. For the analysis of digital library using the digital library websites Gunadarma University, Diponegoro University, University of Indonesia, Bogor Agriculture Institute and the Institute of Technology Surabaya. In some existing digital library on the websites of the digital library is still no automation on his widely read book recommendations based on testimonials from readers of the book is done by looking at the semantic similarity. So that needs to be developed Smart Digital Library using semantic similarity to his widely read book recommendations so that referrals can be more free in expressing their views through a series of words forming testimony. Early stage to find out that the need for a recommendation based on testimony is by distributing questionnaires. Questionnaire deployment was conducted to see the user needs of the features that will be developed at this Smart Digital Library. Preparation of the questionnaire using WebQual and results of data processing using QFD.

2. Research objectives

Based on the issues raised, the purpose of this study is as follows:
1. To identify Consumer Requirements and Technical Requirements that are part of the Quality Function Deployment analysis as applied to Digital Library websites Gunadarma University, Diponegoro University, University of Indonesia, Bogor Agriculture Institute and the Institute of Technology Surabaya.
2. To find out what are the attributes that still require attention based on Quality Function Deployment analysis which has to be used and need to be eliminated in order to obtain a Digital Library applications that suit the needs of the user.

3. Literature review

3.1 Digital Library

Understanding digital libraries or Digital Library there are various opinions. Among that opinion is: as said by Hasibuan [1], a digital library or a digital library system is the concept of using the internet and information technology in library management. Meanwhile, according to Fahmi [2] says that the digital library is a system that consists of hardware (hardware) and software (software), strongly tipped electronic, management staff, users, organizations, working mechanism, as well as services by utilizing various types of information technology, Digital libraries can be grouped into several types as follows:

1. *Early digital libraries*
2. *Digital libraries of institusional publication*
3. *Digital libraries developments at national libraries*
4. *Digital libraries at Universities*
5. *Digital libraries of special materials*
6. *Digital libraries as research project*
7. *Digital libraries as Hybrid library project [3]*

Based on the definitions above that the digital library has basically 3 main characteristics as reviewed [4], namely:

1. Using technology that integrates the ability to create, find, and use information in various forms in a widespread network.
2. It has a collection that includes the data and metadata that mutual linking of data, both internal and external environment.
3. Is an activity to collect and organize digital resources developed together community service users to meet the information needs them.For the digital library is an integration of various institutions that select, collect, process, maintain, and provide information extends to various communities.

Some Reasons of the importance of digital libraries as follows:

1. *A digital Library bring information to the user*
   The digital library can be accessed anywhere. Digital libraries that deliver or bring the information to users whenever they need.
2. *Improved searching and manipulation of information*

The digital library offers a variety of ways of search and retrieval by providing electronic data bases making it easier for users to access information.

3. *Improved facilities for information sharing*
   With digital collections, libraries provide convenience facilities for sharing information between users and between libraries. The digital library also opens opportunities and paves the way for cooperation with other libraries.
4. *Timely access to information*
   The digital library help users get the latest information. The digital library allows to easily access informasiberseri (periodical colection) with digital publishing.
5. *Improved use of information*
   The digital library no longer restricted by time, place, language, and culture, so it's easier use of the information. Diverse information from various parts of the world with different languages and different cultures facilitate search.
6. *Improved collaboration*
   Research at the University of California shows that the digital library facilities capable of improving cooperation among users. This process will improve the dissemination and use of information.
7. *Reduction of the digital divide*
   The presence of the Internet in the world have raised their gab / gap among nations in the world in terms of infrastructure, facilities and resources. The presence of a digital library that can minimize the gap. [3].

Some of the gains of digitalization is one of them is the speed sourcing. Library library should integrate the concept pencarian.Sehubungan with that libraries should take notice 6 (six) of the agreement as reviewed by Tedd and Large [4], namely:

1. Tecnical interoperability (technical agreements), the similarity in the use of procedures and mechanisms of hardware, software, communication protocols, transport of data, procedures for storage and indexing, and others.
2. Semantic interoperability (semantic agreement), the standard use of the term in indexing and retrieval.
3. Political / human interoperability (political agreement): the decision to share together and cooperate.
4. Intercomunity interoperability (agreement among the community of users): an agreement to assemble between institutions and diverse disciplines.
5. Legal interoperability (legal agreement) legislation on access to digital collections, including the matter of intellectual property rights.
6. International interoperability (International Agreements) standard that allows international cooperation, the possibility of other state agencies have the specifications, procedures, technical and legal. [5].

3.2 Quality Function Deployment (QFD)

Quality Function Deployment (QFD) was first developed in Japan in 1966 by Dr. Yoji Akao. Definition of QFD itself according to Dr. Yoji Akao is a method for transforming a request from the user into a quality design for spreading 'function forming quality' and deploy methods for achieving the design quality into the system, component parts, and the specific elements in the manufacturing process. Quality Function Deployment (QFD) is a systematic approach that determines the demand or consumer demand then accurately translate these demands into technical, manufacturing, and production planning right. With QFD, the company's operations is driven by the 'voice of the customer' and not by management commands or opinion / wishes of the experts. The use of QFD focuses on the major causes of satisfaction as well as customer dissatisfaction, making it a useful tool for competitive analysis of product quality by management.

Quality Function Deployment (QFD) is a method of planning and development of products/services in a structured way that allows the development team clearly define needs and expectations and to evaluate the ability of a product or service in a systematic way to meet the needs and expectations [6]. QFD is a structured process or mechanism for determining customer needs and translating those needs into relevant technical requirements, which each functional area and level of the organization can understand and act [7]. Quality Function Deployment (QFD) is a method of planning and development in a structured way that allows the development team clearly define customer needs and expectations, and evaluating the ability of a product or service in a systematic way to meet the needs and expectations [8]. QFD is used to improve understanding of the customer and to develop products, services and processes in a way that is more oriented to the customer [9].

Based on the above definition, it can be concluded that the practice of QFD is to design a process in response to customer needs. QFD translate what the customer needs into what is produced by the organization. QFD enables organizations to prioritize the needs of customers, finding innovative responses to those needs and improve processes to achieve maximum effectiveness. QFD is also a leading practice process improvements that can enable organizations to exceed customer expectations. By definition, QFD is a practice to design a process in response to customer needs. QFD translates what the customer needs to be what the resulting organization. QFD enables organizations to prioritize the needs of customers, finding innovative responses to those needs, and improve processes to achieve maximum effectiveness. QFD is also a leading practice process improvements that can enable organizations to exceed customer expectations. The purpose of QFD themselves as much as possible not only meet customer expectations, but also to go beyond customer expectations as a way to compete, so expect consumers do not resist and did not complain but instead want it.

The main tool of QFD is a matrix [10], where the results achieved through the use of inter-departmental or functional teams to collect, Interpret, document and prioritize customer needs. The starting point (starting point) QFD is a customer and the wishes and needs of customers. In QFD this is called the "voice of the customer" (voice of the customer). QFD is the job of the team heard the voice of the customer. There are three main benefits obtained by the company when using QFD that is: 1. Reduce Costs: This can happen because the products are produced strictly according to customer needs and expectations of consumers so that there is no repetition of work and disposal of raw materials that do not conform to the specifications set by the consumer. Cost reduction can be achieved by a reduction in the purchase cost of raw materials, overheads or wage reduction and simplification of the production process. 2. Increase Revenue: With cost reductions, for the results that we receive will be increased. By QFD products or services produced will be better able to meet customer needs and expectations. 3. Reduce Production Time: QFD will make the product or service development team to focus on program development needs and expectations of consumers [8].

Other benefits derived from the application of this QFD also includes [11]:

a. Focus on the customer (Customer focused) is to get input and feedback from customers regarding the needs and expectations of customers. This is important, because the performance of an organization can not be separated from customers.
b. Efficient time (Time Efficient), by applying QFD, the development program will focus on the expectations and needs of customers.
c. Orientation cooperation (Cooperations Oriented), QFD using a group-oriented approach. All decisions are based on consensus and the involvement of everyone in the discussion and decision-making.

d. Orientation in the documentation (Documentation Oriented), QFD using data and documentation that shows the process of getting all the needs and expectations of customers. Data and documentation is used as information about the needs and expectations of customers are always repaired from time to time.

4.Methods

4.1 Research Design

This research is a case study, namely a depth and contextual analysis of the situation similar in other organizations, where the nature and definition of the problem that occurs is similar to that experienced in the current situation [12]. The data source preformance this study are derived from primary data and secondary data. In the analysis phase of the digital library web sites of data obtained from a digital library the University of Diponegoro (Diponegoro University), University of Indonesia (UI), Bogor Institute of Agriculture, etc.

4.2. Population, sample and sampling technique.

4.2.1. Population

The population is the entire group of people, events, things or objects of interest of researchers who want to investigate [12]. The population in this study is a digital library or digital library.

4.2.2. Sample

The sample is a portion of the population who want to study their characteristics and can be considered representative of the overall population [12]. The sample in this study is part of a digital library on the university digital library among which; library.gunadarma.ac.id, digilib.its.ac.id, digilib.unm.ac.id, digilib.uns.ac.id, digilib.unair.ac.id, digilib.ub.ac.id, digilib.unimed.ac.id, digilib.mercubuana.ac.id, lain-lain.

1.2.3. Sampling techniques

The sampling technique used in this study is simple random sampling. The sample size for this study is estimated at approximately 200 respondents by reason of the determination of the amount based on the opinions [13], that the sampling for the population that is not infinite and the unknown can be taken a sample of 100 people assuming normally distributed population. The total sample of 200 is reinforced by the opinions of Roscoe [12] which states that the number of samples is greater than 30 and less than 500 in most research is represented. Sampling was conducted using questionnaires distributed to users of digital libraries by simple random sampling. This study tries to analyze the quality of service Website Digital Library in order to determine the difference between the actual perception and ideal expectations as measured by the website users WebQual method approach. Instruments prepared by modifying and adding dimension to the three main dimensions, namely WebQual Information Quality, Service Interaction and Usability.

4.2.4. Data analysis methods

4.2.4.1. Descriptive analysis

Descriptive analysis was conducted to determine and be able to explain the characteristics of the variables examined in a situation or to understand the characteristics of organizations that follow certain common practices [12]. The purpose of this analysis is to give researchers a history or to describe aspects that are relevant to the phenomenon of attention from the perspective of a person, organization, industry orientation, or other [12]. The descriptive analysis in this study is to describe the results of research on the design of smart digital library website which is the processing of the respondents using a questionnaire.

4.2.4.2. *Quality Function Deployment* analysis

Quality function deployment (QFD) is an important tool in product planning and development. It helps improve customer satisfaction and shortening product development lifecycle. Due to the fact that QFD is a team tool and assists group process, geographically distributed team members pose a challenging issue on the common practice and usage of QFD [14]. Quality Function Deployment is defined as a structured methodology that is used in the design and development of products to establish specification needs and desires of consumers, and systematically evaluate the capabilities of the product or service to meet the needs and desires of consumers [15]. Quality Function Deployment using matrix-shaped House of Quality, which is used to describe the needs and expectations of consumers and the company's technical ability to design and produce goods or services according to customer desires. Data were collected both primary and secondary processed through the following stages [16]:

5. Discussion

5.1. Characteristics of respondents

Based on the results of the questionnaire obtained distribution is the Library's digital information search by author, showed that 93% of respondents said require a search feature based on the author's name, while the remaining 7% did not need these features.

From the results of the questionnaire, it was found that respondents who need a search feature information by year published the book as much as 80% of respondents, while the remaining 20% do not need these features.

Comparisons between respondents who need a search feature information based on semantic similarity more requiring as many as 83% of respondents stated need, while the remaining 17% did not require a search feature information based on semantic similarity.

From the results of the questionnaire, it was found that respondents who need a feature space to give testimony, many of which require as many as 67% , while the remaining 33% do not need these features.

Comparisons between respondents who require features that provide information on the rating guide so that users know a good book to read based on the ranking of results than if the data shows that more that require as many as 60% of respondents stated need, while the remaining 40% do not need these features.

5.2. Analysis of the Quality Function Deployment House of Quality

Analysis of the Quality Function Deployment House of Quality has been done and has been through the stages as described in the research methodology, which is the stage of determining the attributes of a product by compiling product attributes based on priority (measured by assigning weights interests) that reflect the things that expected by the consumer / user of the product.

Analysis of Quality Function Deployment with the House of Quality that has been done has been through stages, ie the stage of determining product attributes by arranging product attributes based on priority (measured by assigning weights interest) which reflects the things expected by consumer / user of the product. Phase Project Objective conducted with due regard to the comparison of data performance and relatives importance index (weight factor) of product attributes; then we will be able to see opportunities improvements that can be done and set it as a goal to be met in the product design modification project (project objective). The results of the three stages can be seen in Table 1.

The next stage after the calculations obtained from table 1 is the stage of Engineering Characteristics (Technical Parameters). The next stage is the stage of Interaction Matrix that is the core of QFD. This stage is done by connecting the product attributes with their technical parameters. Such relations may be evaluated for each cell matrix, what kind of relationships that happen: a strong-strong (strong), weak (weak) or nothing to do. Results of phase Interaction Matrix can be seen in Table 2. The purpose of QFD is to measure what variables need to be built because of flaws in the digital library were observed and the results of questionnaires QFD which the user states require that its value is below 1. Based on calculations of the Matrix Interaction phase, it can be seen that there is a technical characteristic that has a value of priority in under 1 to be developed for the needs of the user.

Table 1. Product attributes, product evaluation (process benchmarking) and the stage of the project objective

	Questionnaire	Target	Imp.rate	RII	Weight	Weight(%)
Digital Library for completeness catalog	4	5	1,25	5	6,25	8
Digital Library for Accurate Information	4	5	1,25	5	6,25	8
Digital Library for timely information	4	5	1,25	4	5	7
Digital Library for information by category in the catalog	4	5	1,25	4	5	7
Digital Library to search menu information (books / articles / journals)	4	5	1,25	5	6,25	8
Digital Library to search by author	2	5	2,5	2	5	7
Digital Library to search by year of publication	1	5	5	2	10	14
Digital Library to search for information (books / articles / journals) by keyword	4	5	1,25	4	5	7
Digital Library to search for information (books / articles / journals) based on common sense	1	5	5	1	5	7
Digital Library for space opinion / testimony	1	5	5	1	5	7
Digital Library for ranking information of the book based on testimony	1	5	5	1	5	7
Digital Library for ranking information of the book based on testimony Digital Library to download (books / articles / journals)	4	5	1,25	4	5	7
Digital Library to print (books / articles / journals)	4	5	1,25	4	5	7

The Interaction Matrix phase can be seen in the following :

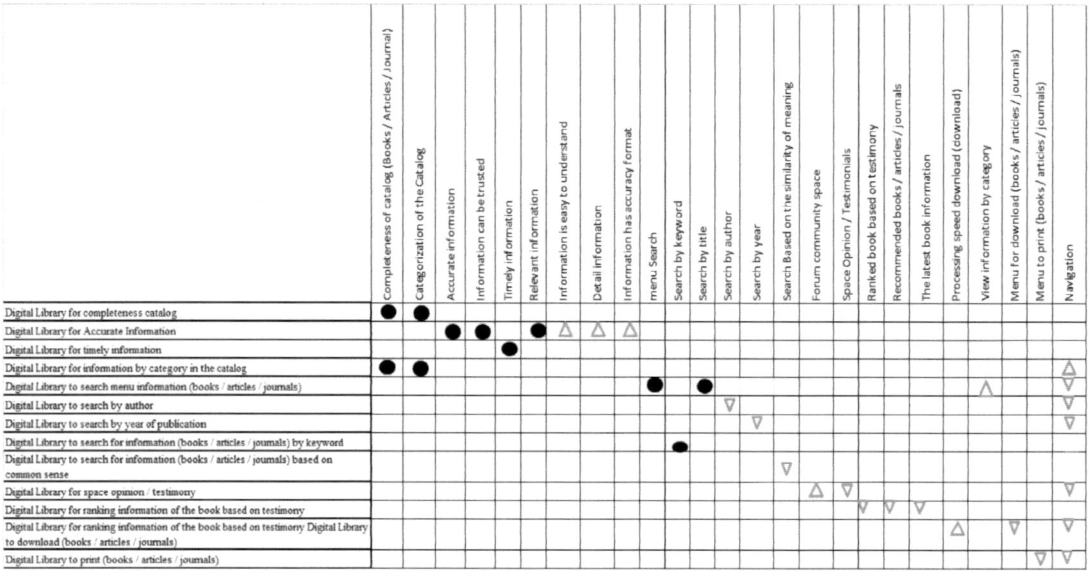

Figure 2. Interaction Matrix

Results of Interaction Matrix phase can be seen in the following table 2.

Table 2. Interaction Matrix Result

No	Engineering Characteristics (Technical Parameters)	Sum Scores	Priority (%)
1	Completeness of catalog (Books / Articles /	135	12,78
2	Categorization of the Catalog	135	12,78
3	Accurate information	72	6,82
4	Information can be trusted	72	6,82
5	Timely information	63	5,97
6	Relevant information	72	6,82
7	Information is easy to understand	24	2,27
8	Detail information	24	2,27
9	Information has accuracy format	24	2,27
10	menu Search	72	6,82
11	Search by keyword	63	5,97
12	Search by title	72	6,82
13	Search by author	7	0,66
14	Search by year	14	1,33
15	Search Based on the similarity of meaning	7	0,66
16	Forum community space	21	1,99
17	Space Opinion / Testimonials	7	0,66
18	Ranked book based on testimony	7	0,66
19	Recommended books / articles / journals	7	0,66
20	The latest book information	7	0,66
21	Processing speed download (download)	21	1,99
22	View information by category	24	2,27
23	Menu for download (books / articles /	7	0,66
24	Menu to print (books / articles / journals)	7	0,66
25	Navigation	92	8,71

Based on the calculation of the Matrix Interaction phase, obtained the technical characteristics that have priority values below 1 to be developed for the needs of users, namely; Search by author, search by similarity of meaning, space Opinion / Testimonials, rating books based on testimony, Recommended Books / articles / journals, information is the latest book, Menu for download (books / articles / journals), Menu to print (books / articles / journal).

6.Conclusions

6.1. The results of data processing using the Quality Function Deployment method has identified Consumer requirements including the need for a digital library. element of consumer needs consist of feature information search by title, name pegarang, and year by considering the similarity of meaning / semantics, room features opinions / testimonials, features books rankings based on the testimony.

6.2. Based on the results of data processing at the Technical Requirements (House of Quality matrix calculation). Stage which is part of Quality Function Deployment analysis, found that there are some design features digital library is not important according to the respondents. Digital design library is designed to be built according to the findings by eliminating features that are not required by the user.

References

[1] Hasibuan, Zainal A, 2005. *Pengembangan Perpustakaan Digital: Studi Kasus Perpustakaan Universitas Indonesia.* Makalah Pelatihan Pengelola Perpustakaan Perguruan Tinggi. Cisarua - Bogor, 17-18 Mei 2005

[2] Fahmi, Ismail, 2004. *Inovasi Jaringan Perpustakaan Digital: Network of Networks (NeONs).* Makalah Seminar dan Workshop Sehari Perpustakaan dan Informasi Universitas Muhammadiyah Malang 4 Oktober 2004.

[3] Chowdhury, G, 2004. *Access to information in digital libraries : users and digital divide.* International Conference on Digital Libraries.

[4] TEDD, L. A., & LARGE, J. A. 2005. *Digital libraries: principles and practice in a global environment.* München, K.G. Saur.

[5] Pendit, Putu Laxman. 2007. *Perpustakaan Digital : Perspektif Perpustakaan Perguruan Tinggi Indonesia* Jakarta: Sagung Seto

[6] Wahyu. Arini, 2003. *Manajemen Kualitas* . Ghalia Indonesia. Jakarta.

[7] Nasution, M. N. 2001. *Manajemen Mutu Terpadu* . Penerbit Ghalia Indonesia, Jakarta.

[8] Ariani, Dorothea W. 2002. *Manajemen Kualitas: Sisi Pendekatan Kualitatif.* Departemen Pendidikan Nasional. Jakarta.

[9] Rampersad, Hubert K. 2006. *Total Performance Scorecard.* Gramedia Pustaka Utama. Jakarta

[10] Alrabghi, Leenah O. 2013,"*QFD In Software Engineering*", Master Thesis, Kent State University

[11] Tjiptono, Fandy & Anastasia Diana. 2003." *Total quality management* ",Andi, Yogyakarta

[12] Sekaran, Uma. 2006,"*Metodologi Penelitian untuk Bisnis*", Salemba Empat, Jakarta

[13] Sugiyono. 2001, " *Metode Penelitian Administrasi* ",Alfabeta, Bandung.

[14] Shah,Smit, Horace Guo, Xun Xu. 2011, "*Development of a web-based quality function deployment system*", International Journal of Internet Manufacturing and Services, Vol. 3, No.1.

[15] Cohen, Louis.1995, *Quality Function Deployment: How to Make QFD Work for You*, Prentice Hall.

[16] Matzler, Kurt and Hans H. Hinterhuber.1998, *"How to make product development projects more successful by integrating Kano's model of customer satisfaction into quality function deployment"*, Technovation, Elsevier Science Ltd.

[17] TEDD, L. A., & LARGE, J. A. 2005. *Digital libraries: principles and practice in a global environment.* München, K.G. Saur.

Institute of Physics
Dirac House, Temple Back
Bristol BS1 6BE UK

ISSN: 1742-6588
ISBN 978-1-5108-3979-3